Water Law SECOND EDITION

Water Law SECOND EDITION

William Goldfarb
Professor of Environmental Law
Cook College, Rutgers University
New Brunswick, New Jersey

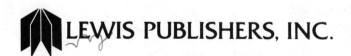
LEWIS PUBLISHERS, INC.

Library of Congress Cataloging-in-Publication Data

Goldfarb, William.
   Water law/William Goldfarb—2nd ed.
     p. cm.
   Includes bibliographies and index.
   1. Water—Law and legislation—United States. 2. Water resources
development—Law and legislation—United States. I. Title.
KF5569.G65   1988   346.7304′691—dc19   87-35553
[347.3064691]
ISBN 0-87371-111-4

Second Printing 1989

COPYRIGHT © 1988 by LEWIS PUBLISHERS, INC.
ALL RIGHTS RESERVED

Neither this book nor any part may be reproduced or transmitted in any
form or by any means, electronic or mechanical, including photocopying,
microfilming, and recording, or by any information storage and retrieval
system, without permission in writing from the publisher.

LEWIS PUBLISHERS, INC.
121 South Main Street, Chelsea, Michigan 48118

PRINTED IN THE UNITED STATES OF AMERICA

346.730 4691
4 618 w
238225

For my parents, Illse, and Catherine

WILLIAM GOLDFARB

A graduate of Yale Law School, William Goldfarb is Professor of Environmental Law at Cook College, Rutgers University, the State University of New Jersey, and is the regular "Litigation and Legislation" columnist and Law Contributing Editor of *Water Resources Bulletin*, journal of the American Water Resources Association.

Goldfarb taught environmental law courses for engineering and science students at Stevens Institute of Technology from 1970 until 1974, when he became Professor of Environmental Law at Rutgers. He now teaches environmental and water law courses to graduate and undergraduate students mainly in engineering, science, planning, and natural resources management. He also participates in multidisciplinary water resources research teams.

Earlier, the author practiced corporate law in New York City, earned a doctorate in English and comparative literature from Columbia University, and taught in that field several years.

As a longtime special consultant to the New Jersey Department of Environmental Protection, he drafted many of the water pollution control laws for that state. He has also served on the New Jersey Governor's Science Advisory Committee and as president of the New Jersey Environmental Lobby.

In addition to the book *Water Law*, Dr. Goldfarb has written numerous articles and book chapters on various aspects of environmental law.

CONTENTS

Acknowledgments   xii

Purpose and Scope of This Book   xiii

INTRODUCTION: Water Law in Context   1
What Do We Mean by Law?   1
Law and Science   4
Uses of Water   6
Water Resource Management   8
What Is Water Law?   9
What Is a Water Right?   10

References for Introduction   13

Part I   THE LAW OF WATER DIVERSION
          AND DISTRIBUTION   15
          1. Legal Classification of Water   17
              Surface Water Categories   17
              Groundwater Categories   18
              Other Categories   19
          2. Water Diversion Doctrines: The Riparian System   21
              What Is Riparian Land?   21
              Which Uses by a Riparian Are Permissible?   22
              Which Uses by a Nonriparian Are Permissible?   24
              Evaluation of Riparianism   24
          3. Water Diversion Doctrines:
              Eastern Permit Systems   26

	Permit and Registration Requirements	26
	Included Waters and Uses	27
	Allocation of Water	28
	Flexibility versus Security	28
	Evaluation of Eastern Permit Systems	29
4.	Water Diversion Doctrines: Prior Appropriation	32
	Appropriationism and Riparianism	33
	Beneficial Use	35
	Water Codes	37
	Classes of Water	39
	Evaluation of Appropriationism	40
5.	Groundwater Doctrines and Problems	42
	Absolute Ownership	43
	Reasonable Use	43
	Restatement Rule	44
	Correlative Rights	44
	Prior Appropriation	45
	Management Systems	45
6.	Federal and Federal-State Diversion Law	49
	Federal Reserved Rights	49
	Federal Regulatory Water Rights	51
	Interstate Waterways	52
	International Waterways	55
7.	Interbasin and Interstate Transfers	56
	Major Interbasin Transfers	56
	State Antiexportation Statutes	58
8.	Drainage Law	61
	Drainage of Diffused Surface Water	61
	Groundwater Drainage	63
	Drainage of Watercourse Overflow (Channel Modifications)	64
9.	Water Distribution Organizations	66
	Municipal Water Supply Organizations	66
	Irrigation Organizations	67
	California's Water Management Districts	68

References for Part I   70

**Part II   WATER RESOURCES DEVELOPMENT
AND PROTECTION**   73

10.	The Federal Government as Project Developer	76
	Bureau of Reclamation	76
	Army Corps of Engineers	79

The Water Resources Development Act of 1986   82
Soil Conservation Service   85
Tennessee Valley Authority   87
Bonneville Power Administration   88
11. The Federal Government as Project Licensor   89
Nuclear Regulatory Commission   89
Federal Energy Regulatory Commission   90
Army Corps of Engineers   93
12. Comprehensive Water Resources
Planning and Research   97
Water Resources Planning   97
Water Resources Research   100
13. Limitations on Federal Development and
Licensing   102
Major Environmental Protection Statutes   103
Other Limiting Statutes and Orders   106

References for Part II   110

Part III   NONTRANSFORMATIONAL USES:
Uses That Do Not Change the Waterbody   113

14. The Public Trust Doctrine   114
15. Public Use of Waterbodies   118
Public Use: Ownership of Beds and Banks   118
Public Access   120
16. Use of Lakes   122
Artificial Lakes   123
17. Use of Rivers and Streams   127
Wild and Scenic Rivers Protection   127
River Corridor Protection   130
Instream Flow Protection   131
Competing Nontransformational Uses   133
18. Federal Outdoor Recreation Programs   135
19. Floodplains Protection   138
The National Flood Insurance Program   139
State Floodplain Regulation   140
20. Wetlands Protection   142
State and Federal Programs   142
Section 404   144
21. Weather Modification   149
22. Acid Precipitation Prevention   153
23. Conservation and Reuse   156
Agricultural Water Conservation   156

　　　　　　Municipal Water Conservation   158
　　　　　　The Supreme Court and Water Conservation   160

References for Part III   161

Part IV   WATER TREATMENT AND LAND USE   165

　　24. The Clean Water Act: Introduction   167
　　25. The Clean Water Act: Goals and Policies   171
　　26. The Clean Water Act: Antidegradation
　　　　　　　and Attainability   174
　　　　　　Antidegradation   174
　　　　　　Attainability   175
　　27. The Clean Water Act: Industrial Point
　　　　　　　Source Dischargers   177
　　　　　　Phase I   178
　　　　　　Phase II   180
　　　　　　Heat Dischargers   181
　　　　　　Prohibited Pollutants   182
　　　　　　Water Quality-Based Effluent Limitations   183
　　　　　　Federal Floors and State Ceilings   188
　　　　　　New Concepts in Industrial Point Source
　　　　　　　Control   188
　　28. The Clean Water Act: Municipal Effluent
　　　　　　　Limitations   190
　　29. The Clean Water Act: Federal Subsidies for
　　　　　　　POTW Construction   193
　　　　　　Construction Grants Program   193
　　　　　　Grants for State Revolving Funds   201
　　　　　　Evaluation of Municipal Point Source
　　　　　　　Control Program   202
　　30. The Clean Water Act: Industrial Pretreatment   203
　　　　　　National Pretreatment Standards   204
　　　　　　Local Pretreatment Programs   207
　　31. The Clean Water Act: National Pollutant
　　　　　　　Discharge Elimination System   208
　　32. The Clean Water Act: Enforcement   214
　　　　　　Governmental Enforcement Remedies   215
　　　　　　Citizen Suits   216
　　　　　　EPA Enforcement   217
　　33. The Clean Water Act: Interstate Water Pollution   219
　　　　　　Interstate Water Pollution Control Compacts   219
　　　　　　Interstate Permitting and Enforcement   221

34. The Clean Water Act: Nonpoint Source Control   223
    Section 208   224
    The 1987 Amendments   228
35. The Clean Water Act: 208 Point Source Planning   231
    Industrial Point Source Planning   231
    POTW Planning   232
36. Ocean Disposal of Waste   233
    Ocean Discharge Criteria   233
    Ocean Dumping   234
37. Groundwater Protection   239
    EPA Groundwater Protection Strategy   240
    Federal Statutory Programs   241
    State Groundwater Protection Programs   248
38. Oil and Hazardous Substances Cleanup   251
    Oil and Hazardous Substances Spills   251
    CERCLA and Hazardous Wastes   253
    State Spill Statutes   256
39. Drinking Water Protection   258
    Safe Drinking Water Act   258
    Saline Water Conversion   262
40. The Common Law and Water Quality   263
41. Regional Land-Use Controls   267
    Adirondack Park Agency   267
    Environmental Land and Water
        Management Act   268
    Lake Tahoe Regional Planning Agency   270
    Watershed Management   271

References for Part IV   272

Part V   MEDIATION OF WATER RESOURCES DISPUTES   275

Glossary of Acronyms   283

Index   285

ACKNOWLEDGMENTS

I would like to express my appreciation to the New Jersey Agricultural Experiment Station, which has provided financial support for my research and writing. Thanks also to Ed Lewis for his courtesy and professionalism; to my colleagues in the Department of Environmental Resources, Cook College, Rutgers, the State University of New Jersey, for their patience and multidisciplinary assistance; to Rich Schiafo for his invaluable aid; and to my graduate students in Water Law for their provocative ideas and esoteric information.

My most heartfelt thanks go out to my wife, Illse Economou Goldfarb, who provided encouragement, support, ideas, and constructive criticism when they were most needed.

PURPOSE AND SCOPE OF THIS BOOK

Law is a critical framework for all water resources decisionmaking. State engineers allocate water based on their state's law of water diversion and distribution. Water quality scientists perform research according to priorities established by law. Planners develop water plans and evaluate proposed projects affecting water resources in the context of federal, state, and local law. Corporate executives order water pollution control equipment to meet requirements contained in their discharge permits. Environmental activists participate in governmental water-related processes under applicable public participation regulations. Consultants submit realistic proposals that presuppose an awareness of relevant legal principles. Whether they realize it or not, all professionals, citizen activists, and students concerned with water resources—be they in the public sector, private sector, academia, or consulting firms—would be more effective if they possessed a fundamental knowledge of water law.

This is especially true because solutions to modern water resources problems are becoming increasingly multidisciplinary and comprehensive. For example, design of a water resources development project might require the participation of engineers, scientists, economists, planners, archeologists, sociologists, political scientists, public administration specialists, and communication experts, as well as lawyers. In order to achieve the optimum result, these individuals should perform as a team, with each team member providing insights and contributing to the formulation of solutions. This process entails a high degree of mutual understanding.

FOR UNDERSTANDING BY ALL DISCIPLINES

Unfortunately, our educational and professional systems overemphasize specialization based on disciplinary orientations. My motive for writing this book has been to counteract the disciplinary myopia that is all too common in water resources education and management. As I remarked in the Preface to the first edition, the most formidable obstacle to rational water resources management is the failure of communication among water resources professionals and among students in the water resources field. Narrow specialization generates esoteric jargon that impairs effective cooperation in pursuit of desirable policy goals. When scientists, engineers, lawyers, planners, social scientists, and management specialists speak intelligibly only to their disciplinary colleagues and students, the results are misunderstanding, chauvinism, and frustration. This book intends to reduce these barriers and to help improve understanding among disciplines.

This book is primarily intended for nonlawyers. Attorneys and law students seeking a broad introduction to water law will also find it useful. But my approach has been to translate and explain legal concepts for people without formal legal training.

Thus, readability has been my guiding principle in writing *Water Law*. Chapters have been kept short and, I hope, to the point. This facilitates using the book either as a text or as a reference work. "Legalese" has been avoided. All legal terms are explained as soon and as thoroughly as possible after their use. In this edition, I have included a Glossary of Acronyms to help the reader digest the "alphabet soup" of nicknames. The Glossary supplements the Index, which is as complete as I could reasonably make it. References have been kept to a minimum, and are placed at the end of each Part where they will be least distracting. Readers who wish to consult the treatises, law review articles, statutes, cases, and administrative regulations cited as references are encouraged to enlist the aid of librarians in their nearest law libraries.

THE SECOND EDITION

Water Law was originally published in 1984. Since then, meaningful trends in water law have intensified, and significant changes in statutory and case law have taken place. For example, both the trend away from riparianism and toward statutory permit systems (administrative allocation) in the East, and the trend toward removing the legal obstacles to agricultural water conservation in the West, have become more pronounced. As for changes in the law, cost-sharing for Corps of Engineers development projects has finally become a reality with the Water Resources Development Act of 1986. In addition, the Clean Water Act, Safe Drinking Water Act, Resource Conservation and Recovery Act, and the Comprehensive Environmental Response, Compensation, and Liability Act have been substantially amended since 1984. In light of these developments, this is an opportune time to introduce a revised and updated edition of *Water Law*.

Although the basic structure of the first edition has been retained, important structural modifications have been made. The Introduction, explaining the nature of law and the relationships among water law, science, water uses, water rights, and water resources management, has been revised and expanded. Material in Part III on recreational use of waterbodies has been rearranged and retitled. Chapters on the public trust doctrine and regional land-use control to safeguard water quality have been added.

Throughout the book, more attention has been devoted to the law relating to groundwater allocation and protection. With all these changes, this edition is considerably longer than its predecessor.

For the Second Printing, a new Part, entitled "Mediation of Water Resources Disputes," has been added. This reflects the recent emergence of mediation as an important mechanism for resolving disputes over water resources.

My labors in writing this volume will be amply rewarded if its availability improves, in whatever small measure, the quality of water resources problem solving in the United States.

WILLIAM GOLDFARB

INTRODUCTION:
Water Law in Context

WHAT DO WE MEAN BY LAW?

For our purposes, "law" is a precept that is written, formally issued by one or more branches of government, and enforceable by governmental sanctions against violators. The function of law is to maintain social cohesion by forestalling disputes—or resolving them once they have arisen—in fair and predictable ways.

Law creates legally enforceable expectations ("rights"), duties to respect those rights, and means of redressing violations of rights ("remedies"). We typically distinguish between "civil" and "criminal" remedies.

Most civil lawsuits ("actions") involve one person "suing" (instituting court proceedings against) another for monetary compensation, "injunction" (a court order to cease an action or perform another), or both. Occasionally, a governmental unit will be the one suing ("plaintiff") or sued ("defendant"). The law treats corporations as "artificial persons" for purposes of suing or being sued. Civil law is invoked when a property owner sues a water polluter for an injunction and compensation for depressed property values ("compensatory damages"), when a pollution control agency sues the polluter for a civil fine, and when a state engineer sues a junior appropriator to enjoin excessive diversions.

In contrast to civil law, which emphasizes financial compensation and cessation of illegal acts, criminal law aims at punishment, deterrence of similar conduct by the violator or others, and rehabilitation. A crime is an illegal act so potentially disruptive to social harmony that, although perpetrated against an individual, it is considered to be an offense against society itself.

1

Thus, governmental units are always plaintiffs in criminal actions. Monetary penalties ("fines") are sometimes exacted in criminal cases, but the most formidable remedy in criminal law is imprisonment. In civil law, on the other hand, imprisonment has realistically been unavailable for over a century. Because of the social stigma and potential deprivation of liberty entailed by a criminal conviction, criminal defendants are almost always entitled to trial by jury; civil juries can only be empaneled where plaintiff seeks damages without an injunction. Moreover, in a criminal case government must prove that a defendant is guilty "beyond a reasonable doubt" in order to obtain a conviction. In a civil case, plaintiff need only prove defendant's liability "by a preponderance of the evidence." Criminal law is becoming increasingly important in water pollution cases where the polluting acts have been heinous, reckless, or repetitive.

There are five major sources of law in the American legal system: (1) constitutions; (2) statutes and ordinances; (3) administrative regulations; (4) executive orders; and (5) common law court decisions ("cases"). Each of these has contributed to the development of water law.

Constitutions are referred to as "fundamental law," meaning that they contain basic principles of justice that underlie other legal principles and procedures. There are a Constitution of the United States, applicable to all American citizens, and state constitutions that afford rights only to citizens of particular states. The federal Constitution and its state counterparts are similar but not identical. For example, like the federal Constitution, most state constitutions contain "takings clauses," forbidding the taking of private property for public use without just compensation. It will be seen that the definition of "taking" is central to a determination of whether, and to what extent, government may legally regulate water-related land use. An unusual provision is the clause in some western state constitutions that all unappropriated surface waters are available for appropriation. Because constitutions are generally phrased in broad terms, the role of courts as final constitutional interpreters is a critical one. State supreme courts, with regard to state constitutions, and the United States Supreme Court, where the federal Constitution is concerned, are responsible for applying ambiguous constitutional terms such as "taking" and striking down statutes and regulations that violate constitutions (e.g., that are unconstitutional takings without just compensation).

Statutes, or "acts," are bills that are passed ("enacted") by appropriate votes of state legislatures or Congress. On the county and municipal levels, legislative enactments are referred to as "ordinances." However, a statute does not achieve the force of law until it is signed by the president or governor, as the case may be, or until the chief executive's veto is overridden by the required legislative vote. Both the federal Clean Water Act of 1972 and the Water Quality Act of 1987 became law when Congress overrode a presidential veto. In an increasing number of states, water law is primarily state statutory law. Although courts interpret ambiguous statutes in accordance with their conceptions of legislative intent, a legislature can always overturn a judicial interpretation by amending the statute, as long as the chief executive agrees or sufficient support can be mustered to override a veto. Courts have the final say regarding the meaning of constitutions, but legislatures are the final interpreters of statutes. Courts are also important in enforcing statutes. For example, violators of the Clean Water Act may receive judicially imposed civil penalties of up to $25,000 per day of violation or criminal penalties of up to $250,000 and 15 years in prison.

Because of the need for continuous expert regulation in highly technical areas such as water pollution control and water distribution, Congress and state legislatures enact statutes creating administrative agencies. An "enabling act" or "organic act" establishes a regulatory framework, sets out general rules and standards for the agency to follow, and delegates to the agency the authority to issue rules and regulations carrying the force of law. Thus, where the agency has developed and issued ("promulgated") a rule or regulation in accordance with the applicable administrative procedure act, the promulgated rule can be enforced against violators either by the agency (for example, through an administrative fine) or by a court at an agency's behest. In contrast, administrative "guidelines," issued without the required public notice and comment, are not "law" because they are not directly enforceable. Administrative agencies also "make law" when they issue written decisions adjudicating disputes over licenses or permits; for example, when there are contested claims to priority of diversion in prior appropriation states. Administrative agencies are governmental microcosms in that they exercise legislative powers (i.e., rule making), judicial powers (i.e., adjudication of conflicting claims), and of course, executive powers. The major federal water law agencies are the United States Department of the Interior (DOI), the Environmental Protection Agency (EPA), and the National Oceanic and Atmospheric Administration (NOAA) in the United States Department of Commerce. At the state level, water law activity is centered in state environmental protection agencies and "state engineer" offices.

The power to issue executive orders is inherent in the chief executive's authority to supervise the executive branch and deal with emergency situations. In 1970 President Nixon created the EPA by reorganization plan, transferring powers from the DOI, the United States Department of Agriculture (USDA), and the Department of Health, Education and Welfare (now the Department of Health and Human Services). Some years ago the Governor of New Jersey ordered his Department of Environmental Protection to use existing permitting authority to curtail development in the Pinelands in order to protect a massive and vulnerable aquifer from contamination by septic tanks. Executive orders have the force of law, but important legal questions regarding them remain unanswered. What is the legal effect of an executive order on a legislature? Can a private citizen sue to enforce an executive order? Most important, what is the permissible scope of an executive order?

The common law frequently comes into play when a property owner sues a neighbor for polluting or diverting water that the plaintiff is entitled to use. Preventing a neighbor's drainage water from damaging property is another kind of common law action. Common law is present when a lawsuit is designated as based, for example, on nuisance, trespass, negligence, violation of drainage rights, violation of riparian rights, or interference with the reasonable use of groundwater. Unlike statutory law, common law is judge-made law. It is contained in written judicial decisions in specific lawsuits. These decisions are known as "cases." The keynote of common law is precedent—in similar factual situations judges are expected to follow the legal rules established by their predecessors. An excessive devotion to precedent at the expense of fairness is avoided by encouraging judges to distinguish situations factually and by recognizing the power of a state supreme court or legislature to change common law rules. For example, a number of states have judicially substituted the "reasonable use rule" for the "common enemy rule" in drainage cases. Although the common law is of limited importance in water quality law, it remains viable in drainage and water quantity disputes, especially in the East. The availability of common law actions supports statutory water law by bringing novel situations and their judicial resolutions to the attention of legislatures. Common law also offers an individual the ability to recover "compensatory damages" for property damage and personal injury. Most statutes do not offer individuals the right to recover the cash value of personal loss.

LAW AND SCIENCE

Law and science are different conceptual systems. They have different goals, different notions of truth, and different methods of proof.[1] Scientists are often amazed that "good science" may not prevail in court. For example, science has statistically proved that cigarettes cause lung cancer, but common law courts do not accept this conclusion because they do not accept statistical evidence. Causation in common law means "proximate cause": plaintiff must prove, by a preponderance of the evidence, that his injury would not have occurred without defendant's action or inaction, and that his injury must have been foreseeable to defendant. These demands place an awesome burden on a plaintiff where scientific information is incomplete.

However, not all lawmaking bodies are frustrated by the common law's conservative view of causation. Legislatures and administrative agencies are responsible for protecting public health, safety, and welfare. They are permitted to err on the side of safety by enacting statutes and promulgating regulations as long as these have been formulated rationally, based on scientifically arguable data. Legislatures and agencies need not wait for scientific certainty when there is potential danger to the public. Their lawmaking will be upheld where it is not "arbitrary"; that is, where it is reasonable, based on the perceived public threat and on the available scientific data.

Legislating in advance of scientific certainty has its pitfalls, however. Indeed, many important water law principles are based on obsolete science. For example, one of the most prevalent distinctions in water law is that between surface water and groundwater, a distinction that is minimized by modern hydrology. In fact, many states apply different legal principles to surface water and groundwater. Moreover, in some of these states the applicable groundwater rule is based on the assumption that groundwater moves in fundamentally inscrutable ways. How can the law be so retrograde in lagging behind science?

One answer is that legal decisionmaking is flexible enough so that rules based on outdated science do not necessarily result in bad decisions. Lawmakers find ways to resolve controversies equitably, even if it requires sophistry. Then, at the point where the legal stable becomes too Augean with untenable rules, conflicting rules, and distorted precedents, a supreme court or legislature will intervene and wash away the solid waste.

Just such a judicial reconciliation between law and science may be found in a 1982 decision of the Rhode Island Supreme Court, entitled *Wood* v. *Picillo*.[2]

This case involved a common law "nuisance" action by neighbors, a cranberry grower, and members of the public against a farmer who maintained a hazardous waste dump on his property. Plaintiffs alleged that defendant's disposal activities emitted noxious fumes and polluted ground and surface waters. According to defendant, plaintiffs bore the burden of proving that he could reasonably have foreseen that his actions would pollute groundwater and, as a result, downgrade surface water. Defendant's argument was based on a 1934 decision entitled *Rose* v. *Socony-Vacuum Corp.*, where the same Rhode Island Supreme Court, in holding for defendant, had reasoned that because "courses of subterranean waters are indefinite and obscure," it would be unjust to subject landowners to liability unless plaintiffs could prove that they should have foreseen the consequences of their actions (i.e., that defendants had been "negligent").

In overruling the *Rose* decision and holding for plaintiffs, the court stated,

> Since this court decided *Rose* v. *Socony-Vacuum* in 1934, the science of groundwater hydrology as well as societal concern for environmental protection has developed dramatically. As a matter of scientific fact the course of subterranean waters are no longer obscure and mysterious. The testimony of the scientific experts in this case clearly illustrates the accuracy with which scientists can

determine the paths of groundwater flow. Moreover, decades of unrestricted emptying of industrial effluent into the earth's atmosphere and waterways has rendered oceans, lakes, and rivers unfit for swimming and fishing, rain acidic, and air unhealthy. Concern for the preservation of an often precarious ecological balance, impelled by the spectre of 'a silent spring,' has today reached a zenith of intense significance. Thus, the scientific and policy considerations that impelled the Rose result are no longer valid. We now hold that negligence is not a necessary element of a nuisance case involving contamination of public or private waters by pollutants percolating through the soil and traveling underground routes.

At its best, law is a dynamic process that remains as consonant with science as possible in light of the fundamental differences between the two systems.

USES OF WATER

Developments in science and technology profoundly influence our uses of water. This statement is self-evident as regards the first seven of the nine traditional water uses: navigation, irrigation, potable water, waste disposal, industrial process and cooling purposes, hydroelectric power, and recreation on the water. However, it is also true for fish and wildlife maintenance and scenic/aesthetic enjoyment. We can breed or transplant strains of fish that are resistant to pollution, water temperature fluctuations, and increased fishing pressure. In addition, television has fostered an enhanced appreciation of scenic beauty and helped to create a political constituency for preservation.

Conversely, our need for water affects the development of science and technology. Water shortages stimulate research into novel methods of water production, storage, and distribution. Public demand for clean water prompts industry to invest in new processes that recycle water.

Like science, water law both influences and is influenced by our uses of water. Safe drinking water legislation has compelled water purveyors to upgrade treatment and raise water rates. The proliferation of faster speedboats has necessitated speed limit regulations to prevent accidents. Thus, the relationship between water law and water use is a reciprocal and dynamic one. This illustrates the law's tendency to both frame and respond to those controversial matters that we refer to as "issues." However, this relationship has been distorted by our neglect of a major water use—conservation. We have historically treated conservation (i.e., not supply-side conservation but demand-side conservation) as an afterthought, a set of undesirable measures to be imposed only where there is insufficient water for human use. But times are changing. Water is, and will continue to be, scarce in many parts of the United States. Consequently, we are learning to perceive water conservation as an end in itself, a use on a par with other uses of water.

Recognition of intentional nonuse and use reduction as a legitimate water use will have significant impacts on water law, especially in the West where legal rules regarding water demand conservation are few and relatively primitive. We can confidently expect water conservation law to be one of the most volatile areas of water law for the foreseeable future.

There is another basic flaw in our traditional analysis of water uses, which is illustrated by this excerpt from *Water Policies for the Future*, the final report of the National Water Commission (NWC) (1973):

> Water use consists of (1) intake uses, (2) onsite uses, and (3) instream or flow uses. Intake uses include water for domestic, agricultural, and industrial purposes—-uses that actually remove water from its source. Onsite uses consist mainly of water consumed by swamps, wetlands, evaporation from the surface of water bodies, natural vegetation, unirrigated crops, and wildlife. Flow uses include water for estuaries, navigation, waste dilution, hydroelectric power, and some fish and wildlife and recreational uses.
>
> Water uses are measured in two ways, by amount withdrawn and amount consumed. Water withdrawn is water diverted from its natural course for use, and may be returned later for further use. Water consumed is water that is incorporated into a product or lost to the atmosphere through evaporation and transpiration, and cannot be reused. Water consumption is the more important indicator, since some part of withdrawn water can usually be reused, although not always near the point where the first withdrawal takes place.

This formulation is seriously flawed, because, as was characteristic of water resources management in the past, it is oriented toward water quantity, not water quality. Use of water for waste removal partakes of all three uses: a cooling tower is an intake use, uptake of nutrients by a wetland is an onsite use, and BOD (biochemical oxygen demand) assimilation is an instream use. Moreover, in terms of water quality, it is unproductive to place hydroelectric power generation in the same category as scenic preservation. As for measuring water uses, the distinction between "withdrawal" and "consumption," as these terms are defined by the NWC, does not help us to evaluate waste discharges or potential development. Pollution and destruction of fish and wildlife or scenic beauty by water resource development are in a very real sense consumptive uses, although water may not be physically lost to the atmosphere.

Instead of redefining "consumptive use" in a tortured fashion inconsistent with tradition and everyday speech, entirely new phraseology is introduced where appropriate in this book. Water uses are spoken of as "transformational" or "nontransformational." A "transformational use" is one which causes a significant change in the existing condition of a waterbody. This includes substantial diversion or waste discharge as well as heavy recreational use. A "nontransformational use" leaves a waterbody in a basically unaltered state. This approach has the advantage of placing all water uses on the same plane. Thus, it deals directly with the fundamental value question inherent in water resources management: Are the social and economic gains produced by water use worth the costs of ecosystem change?

This more balanced view of water use will continue to transform water law. Protecting human health and the environment from the adverse impacts of water resource development and related land use has become an ineradicable priority of our legal system.

WATER RESOURCE MANAGEMENT

Water is a primary natural resource, in the sense of being necessary for human existence and frequently in short supply. In order to be useful to people, a natural resource must be available in adequate quantity, in appropriate quality, in the right place, at the right time, and at reasonable cost. Water resource management is the maximization of as many of these variables as possible in response to demands—frequently, conflicting demands—for water. Water resource policy (or public policy) consists of those public decisions, and the criteria on which they are made, that affect water resource management. Law is one means of formulating and effectuating water resource policy (electoral politics, the marketplace, and the media are some others). A well-spacing statute, for example, defines the rights and obligations of administrative agencies, well drillers, and water users such as homeowners and farmers. In addition, the statute provides mechanisms for administering the policy (e.g., permits for well drilling) and enforcing it, should it be violated.

Water resource management should be comprehensive and holistic. However, our historical separation of water quantity and water quality has impeded the development of an effective water resource management and its concomitant water law. A Conservation Foundation Report points out that

> (t)he United States generally has dealt with its two major types of water problems—water quantity and water quality—in different statutes, by different agencies, and at different levels of government. Yet the problems are closely related and solutions dealing with only one problem often end up exacerbating the other.[3]

The report illustrates the interaction between quantity and quality by three examples involving salinity. Progressively higher levels of salinity in the Colorado River have been "caused less by greater additions of salt than by consumption of the river's fresh water—water that formerly diluted the salt, keeping its concentration down." Law, responding to public policy, has both encouraged Colorado River diversions and facilitated construction of a $250 million desalting plant in Yuma, Arizona, to meet treaty obligations with Mexico to deliver 1.5 million acre-feet of water per annum with acceptable salinity. In North Carolina, "the drainage of . . . coastal wetlands for farming, aided by federal and state agencies interested only in agricultural productivity, has been reducing the salinity of one of the largest estuarine systems on the Atlantic Coast," to the detriment of a profitable fishing industry. In addition, many aquifers are becoming useless for potable water because of saline intrusion caused by groundwater mining.

Integration of water quantity and quality is a recurring theme in modern water resource management and water law, a theme that will be explored more thoroughly in subsequent parts of this book.

WHAT IS WATER LAW?

Most writers on water law declare that "water law is the creation, allocation, and administration of water rights."[4] But this is an inaccurate, unproductive definition based on the artificial dichotomy between water quantity and quality.

Traditionally, the general term "water law" has been synonymous with legal rights to particular amounts of water, especially in the American West. Casebooks and commentaries on water law frequently also discuss matters related to water quantity, such as federal water resource development and public recreational access and use. However, the law of water quality is almost always omitted from, or given short shrift by, these books. For example, a recently published casebook entitled *Legal Control of Water Resources* devotes only 50 desultory pages of its 934 total pages to various aspects of water pollution control law.[5] A modern concise commentary on water law dispenses with water pollution in seven pages.[6] Indeed, "water quality law" is typically viewed as a distinct field of law, separate from "water law." In other words, the particular term "water quantity law" has swallowed the general term "water law." This use of terminology is obsolete in an era when we recognize the need for comprehensive water resource management characterized by the interrelating of hitherto disparate categories such as water quantity and quality. Nowadays, water law should encompass the legal aspects of all water uses.

Like most fields of law, "water law" defies precise definition. It overlaps other legal fields such as environmental law, natural resources law, property law, tort law, public land law, and land use law. Nevertheless, "water law" is loosely defined here as those aspects of law that are of

primary concern in the management of water resources. For analytical purposes, the field is subdivided as follows: (1) the law of water diversion and
distribution, (2) the law of water resource development and protection, (3)
the law of nontransformational uses, (4) the law of water treatment, and
(5) the law of water-related land use. Wherever possible, relationships among
these subdivisions will be elicited.

Because of space limitations, another aspect of water law; namely,
oceans law and international water law, is incompletely treated here. These
issues are discussed only where they clarify the direct focus of this book,
which is the law applicable to fresh and tidal waters of the United States,
and to those waterbodies the United States shares with Canada and Mexico.

Even within these limitations, it is clear that the multiplicity of American water resource legal issues cannot be thoroughly analyzed in a single
volume. For example, the legal regime applicable to the Colorado River and
the legal problems relating to water use in the Colorado Basin deserve a
book to themselves. Instead, this book emphasizes general water law principles and attempts to provide references for those readers who desire to
study the legal aspects of local or regional water resources problems in greater
detail.

WHAT IS A WATER RIGHT?

Water law deals with water rights of various kinds. A legal right is a legally
enforceable expectation—the assurance that an activity will be protected by
the legal system. It means that other people have enforceable duties toward
a rightholder.

Legal rights are either private or public. A set of legal rights that a rightholder holds exclusively ("owns") is called "private property." Ownership
of private property defines a group of rights with regard to the subject of
property; e.g., a plot of land or an automobile. These property rights generally include the right to possess the subject, use it, dispose of it, and prohibit anyone else from interfering with it for an indefinite period of time.

Public rights are held in common by a definable group of people. The
rights guaranteed by the United States Constitution are the heritage of all
American citizens. Each member of the class of rightholders may assert a
public right individually, even though the right is common to all. Thus, each
American citizen has legal authorization ("standing") to sue a governmental unit if it frustrates his exercise of a fundamental right such as freedom
of speech.

No right—private or public—is absolute. Landownership is subject to
the public rights of other community members as expressed in zoning and

other regulations. The constitutional right to freedom of speech does not permit unjustifiably crying "fire" in a crowded movie theatre. Thus, all rights exist in a delicate state of equilibrium with other rights.

When private and public rights conflict, one mediating mechanism is constitutional "takings clauses." Under the Fifth Amendment of the United States Constitution and its counterparts in state constitutions, no government is entitled to "take" private property for public use without paying "just compensation." Takings law applies not only to the usual "condemnation" situation where government locates a right-of-way through private property, but also where regulations excessively burden private property. In either case, government must pay a private property owner the fair market value of the lost property interest.

One of the thorniest problems perplexing the American legal system is: At what point does a governmental regulation to protect the public health, safety, and welfare become a "regulatory taking" of private property for which just compensation must be paid? This issue will be explored more thoroughly in later chapters, but a few extreme examples here will illustrate the concept.

Water quantity rights may be subject to reasonable public regulation such as rules governing registration or forfeiture for nonuse. But a water quantity right is private property and government cannot usurp a water quantity right without paying fair market value for it. In a number of western states, water-short municipalities are statutorily authorized to take over senior agricultural diversion rights, but only after payment of just compensation to farmers and ranchers.

The public has a right to recreational opportunities, but this does not justify zoning private lakefront property as a public waterfront park. If government wants to create a park on that site, it must purchase the land or public access rights from the owner.

A right to a particular quantity of water is private property because it is unique to the rightholder, but it is a different mode of property than that in a plot of land or an automobile. Water in a natural waterbody is so important to society that it cannot be privately owned. It is owned by each state as trustee for its citizens. In other words, water in natural waterbodies is "public property"; it belongs to all citizens of the state. A private citizen can only own the right to use such water. This is what lawyers call a "usufructuary right." It is not meaningless semantics to say that the subject of a private property right in water quantity is not the water itself but the use of water, even though in many American states water rights can be bought and sold. For one thing, the distinction supports extraordinary state powers to regulate water use, powers which a state can assert against the federal government as well as its own citizens. After all, as true owner of the water in natural waterbodies within its boundaries, a state should possess greater authority to regulate water use than to regulate land or automobile use. But if government goes too far in protecting public rights at the expense of private rights, its actions may be declared "takings" requiring the payment of compensation.

REFERENCES FOR INTRODUCTION

1. Large, D. W., and P. Michie. "Proving That the Strength of the British Navy Depends on the Number of Old Maids in England: A Comparison of Scientific Proof with Legal Proof," *Environ. Law* 11:555–638 (1981).
2. 433 A. 2d 1244 (1982).
3. *State of the Environment: An Assessment at Mid-Decade* (Washington, DC: Conservation Foundation, 1984), p. 373.
4. Rice, L., and M. D. White. *Engineering Aspects of Water Law* (New York: John Wiley & Sons, 1987), p. 1.
5. Sax, J. L., and R. H. Abrams, *Legal Control of Water Resources* (St. Paul, MN: West Publishing Co., 1986).
6. Getches, D. H. *Water Law in a Nutshell* (St. Paul, MN: West Publishing Co., 1984).

PART I

The Law of Water Diversion and Distribution

The American legal system is not monolithic. In fact, there are hundreds of major legal systems, if those of the federal government, the 50 states, and the largest cities and counties are included. The law of water diversion and distribution is predominantly state law. These rights to use water are almost invariably declared in state constitutions and statutes, adjudicated by state courts and administrative agencies, and administered by state agencies. Since states are independent sovereigns in the water law field, different states, even neighboring states, often apply contrasting legal principles to a particular water law problem. Consequently, in water law we regularly refer to "majority" or "minority" state rules, depending on the number of states adhering to that rule.

Where water diversion and distribution are concerned, state law generally reigns supreme unless a state oversteps its bounds and violates a right embodied in federal law. For example, rights of irrigators to water developed by federal reclamation projects are governed by federal statutory reclamation law. State law in conflict with this federal law is null and void. Federal claims to reserved water rights are decided under federal law, even though they may actually be adjudicated in state courts. Moreover, because a water right is treated by the law as "property," under the Fifth Amendment to the United States Constitution no state may "take" a water right for public use without "just compensation" to the rightholder. But these exceptions only prove the rule of state hegemony in matters of diversion and distribution. Indeed, the federal government has always paid extraordinary deference to state water law in these areas.

Chapter 1

LEGAL CLASSIFICATION OF WATER

Over hundreds of years, and without reliable hydrologic data, courts and legislatures have divided the hydrologic system into numerous classes of waters that are subject to different legal rules. The trend is away from this unscientific set of distinctions, but legal change occurs slowly, especially in a field, such as water diversion and distribution law, that is a function of 50 different state legal systems and federal law.

The fundamental legal classifications of water are groundwater and surface water. Nowadays, when enlightened water resource management requires conjunctive use of waters based on the interconnections of groundwater and surface water, this distinction is a particularly harmful one. The following is a brief introduction to the legal classes of groundwater and surface water.

SURFACE WATER CATEGORIES

Diffused Surface Water

Diffused surface water is the uncollected flow from falling rain or melting snow, or is spring water that spreads over the earth's surface.[1] It follows no defined course or channel and forms no more definite body of water than a bog or marsh. Water loses its character as diffused surface water when it reaches a well-defined channel. In law, diffused surface water is analogized to a wild animal and the "law of capture" is applied to it. "Diffused surface water may be captured by the owner of the land over which it moves, and when captured becomes the property of the landowners."[1] Unlike water in watercourses, diffused surface water is almost never subject to state permit programs.[2] The reason for this diffused surface water rule is public policy encouraging agricultural impoundment and use. Most legal problems involving diffused surface water arise when a landowner tries to get rid of it. "Drainage law" is covered in Chapter 8.

Water in Watercourses

The term "watercourse" generally includes all surface waters contained within definite banks. A watercourse may be running water, such as a creek or river, or a body of still or flat water, typically a lake or pond.[3] In some states, floodwaters that have become permanently severed from the main current when waters recede are treated as diffused surface water, capturable by the landowner, and not water in a watercourse, subject to public water rights. Other states consider all floodwater as diffused surface water.[1] Courts are also sharply divided on how to classify springs, categorizing them variously as diffused surface water, groundwater, or part of a watercourse depending on the facts of each case and the legal rule applicable in a specific state. The same is true with regard to the legal classification of marshes and swamps. It is possible that the same hydrologic swamp located in three different states would be classified in three different ways and subjected to three different legal rules.

Courts frequently distinguish watercourses from diffused surface waters by declaring that watercourses are characterized by regular flow and defined channels. But in many states, especially in the West, intermittent streams have been held to be watercourses. The "defined channel" test appears to be a more reliable one. Artificial watercourses (impoundments, artificial channels) are governed by different legal rules than natural watercourses. (See Chapter 16).

GROUNDWATER CATEGORIES

It is inaccurate and confusing to distinguish among subflows (underflows) of surface watercourses, underground streams, and percolating groundwater. But courts continue to make these distinctions, and important legal consequences depend on them.

Subflow of Surface Streams

This category refers to the saturated zone directly beneath and supporting a river or lake in direct contact with surface water. Where subflow can be identified, it is considered as part of the watercourse itself.[4]

Underground Streams

An underground stream is defined as water that passes through or under the surface in a definite channel. In water law, a subterranean stream is treated as a surface watercourse. There is a legal presumption against groundwater being an underground stream; that is, a claimant must produce convincing evidence that underground water flows in a definite and known channel, and does not "percolate" as in an aquifer.[5] This evidence might include surface vegetation along its course, test borings, sounds of the water, geologic data, or interconnections with surface streams.

Percolating Waters

These include all waters that pass through the ground beneath the surface of the earth without a definite channel, and groundwaters that are not shown to be directly connected to surface water. Percolating waters come from precipitation that infiltrates the soil, streamflow from losing streams, irrigation return flow, or artificial recharge. Storage of percolating waters may be in unconfined (watertable) or confined (artesian) aquifers. Rivulets and veins of water within an aquifer are not underground streams but percolating water. In many states, water rights in surface watercourses and underground streams are determined on radically different legal principles than rights in percolating waters.

The state of Colorado makes a further distinction between "tributary" and "nontributary" percolating waters. Groundwater that will eventually reach and become part of a natural stream, either surface or subterranean, is treated as part of that stream for legal purposes (i.e., governed by the rules applied to surface watercourses).[6]

OTHER CATEGORIES

Wastewater

Alternately referred to as "seepage," "leakage," or "artificial water," these are waters that, in the words of a New Mexico statute, are "due to escape, seepage, loss, waste, drainage, or percolation from constructed works, either directly or indirectly, and which depend for their continuance upon the acts of man.'' Seepage from reservoirs and irrigation ditches as well as discharges from treatment works are examples of wastewater. Irrigation return flow (''tailwater'') is a particular kind of wastewater that is sometimes subject to different legal rules. The major diversion and distribution law problem involving wastewater is whether it belongs to the person who releases it or a downstream, or groundwater, rightholder.

Foreign Water

This is water that has been imported by a user from one watershed into another.[1] It will be seen in Chapter 7 that interbasin transfers of water have caused many legal problems. Rights to seepage and return flow from foreign water can also be troublesome.

Storage Water

Impoundment of water is becoming increasingly important for irrigation, recreational, hydroelectric, and municipal uses. Legal problems include liability for negligent storage, the question of storage rights versus direct-flow rights, and the need for environmental impact statements.

Salvage Water

Salvage water is water saved by conservation practices that would otherwise have been lost to seepage or evaporation. Conservation has been impeded in the West by rules that make water saved by conservation subject to use by others. (See Chapters 4 and 23).

Developed Water

Unlike salvage water, which would naturally be part of a watercourse, developed water is ''new water'' made available by human effort. Drainage into a stream of a mine or wetland are examples of developed water.

Chapter 2

WATER DIVERSION DOCTRINES: THE RIPARIAN SYSTEM

The term "riparian," used as an adjective, means "of, pertaining to, or situated or dwelling on the bank of a river or other body of water."[7] Thus we speak of riparian homes or riparian lands. Used as a noun it refers to one who owns land on the bank of a natural watercourse. A riparian right is one held by a riparian, and the riparian system (or riparianism) is the legal apparatus for allocating and enforcing riparian diversion rights. Riparianism is the water rights system that is in effect in all states east of the Mississippi River except Mississippi, and in Arkansas, Iowa, and Missouri to the west.[8] In many of these states, riparianism has been supplemented by permit systems (see Chapter 3). The prior appropriation system prevails in the West (see Chapter 4). However, riparian diversion rights coexist with appropriative rights in those western states that follow the "California Doctrine."

The key tenet of the riparian doctrine is that only persons owning land on natural watercourses possess riparian rights. There can be no riparian rights in groundwater, diffused surface water, or water in artificial waterbodies.[9]

WHAT IS RIPARIAN LAND?

The riparian system not only restricts water diversion rights to riparian owners, but also restricts the use of diverted water to riparian land. In most riparian states a riparian owner, even if he has not actually been harmed, can obtain a court order enjoining another riparian from using water on his own land (1) not touching the waterbody, or (2) outside the watershed.[10] Other states have softened these place-of-use restrictions by requiring that a riparian suffer tangible harm before being allowed to recover damages or obtain an injunction. Still other states allow a riparian owner to make "reasonable" uses of water on unconnected or trans-watershed lands.[10]

There are two legal tests employed to delineate the inland extent of riparian lands. The "source of title" test, adopted in California, limits riparian land to the smallest parcel held in one ownership that, throughout the chain of title, has never been severed from the stream. Thus, from the time of initial government or royal grants, the amount of land possessing riparian rights has diminished each time a tract has been subdivided and sold. Acquisition of adjacent land, although once riparian, does not increase the amount of land benefited by riparian rights.[11]

A more inclusive definition of the extent of riparian land has been embraced in the Southeast. This test, known as the "unity of title" test, generally holds that riparian land extends to any land held in a single title contiguous to a watercourse. Thus the land subject to riparian uses increases as parcels adjacent to existing riparian tracts are acquired, and decreases as off-stream portions of riparian land are sold. Given the trend toward large agricultural landholdings, application of the unity of title test may result in appreciably expanding the quantity of riparian land.[11] As with all aspects of water diversion law, each riparian state's system must be examined carefully to determine which place-of-use restrictions and definition of riparian land are applicable.

WHICH USES BY A RIPARIAN ARE PERMISSIBLE?

Riparian owners in all states—even prior appropriation states—have the following rights, subject to applicable governmental restrictions: (1) the right to have access to the watercourse; (2) the right to fish; (3) the right to wharf out to deeper water; (4) the right to alter or protect shoreline areas; (5) the right to recreate on all or part of the surface (see Part III); and (6) the right to claim title to the beds of nonnavigable watercourses (see Part III).[8,10] These are riparian rights common to owners of land abutting natural watercourses in all states. But the rules of diversion and distribution characteristic of the riparian system are only found in states where riparianism prevails.

As inherited from the common law, riparian diversion rights were originally defined by the "natural flow doctrine": each riparian owner was entitled to have a stream flow through his land in its natural condition, not materially retarded, diminished, or polluted by others.[9] The so-called "natural uses" of water (domestic bathing, drinking, gardening, and household stockwatering) were unlimited. But major "artificial uses" (irrigation, manufacturing, power generation, mining, and stockwatering) were restricted to the mouths of waterways. A lower riparian owner could play "dog in the manager"—not using the water but depriving upstream owners of its use by insisting on the maintenance of natural flow.[9] Because it is inherently antidevelopment, the natural flow doctrine has gradually been replaced by the "reasonable use rule."

Under the reasonable use rule, every riparian owner has a right to divert water for any purpose if the use is reasonable with respect to other riparians, i.e., does not unreasonably interfere with their legitimate uses.[9] Thus, riparian diversion rights are said to be "correlative," mutually interdependent and not merely based on benefit to a single user. Whether a use is reasonable is a question of fact to be resolved by courts on a case-by-case basis. Factors that affect this judicial determination are: (1) the purpose of the use; (2) the suitability of the use to the watercourse; (3) the economic value of the use; (4) the social value of the use; (5) the extent and amount of the harm it causes; (6) the practicality of avoiding the harm by adjusting the use or method of use of one proprietor or the other; (7) the practicality of adjusting the quantity of water used by each proprietor; (8) the protection of existing values of water uses, land, investments and enterprises; and (9) the justice of requiring the user causing harm to bear the loss.[10] Reasonableness is relative, and "in suits between riparian landowners the reasonableness of both uses is an issue."[10] During periods of shortage, some uses may be deemed unreasonable under the circumstances and curtailed. Moreover, priority of use is only one factor among many to be considered by a court. Thus, "in practice as well as in theory, the riparian doctrine generally allows entry of new users and requires old users to accommodate them."[12] For the most part, under the riparian doctrine the relationship among present and future riparian uses on a particular watercourse is one of parity rather than priority.

Nevertheless, there are some "preferred uses" under the riparian system:

> Beyond the general preference for domestic uses, states by legislation have adopted additional preferences among users. Perhaps the most common and the earliest preference favored milling operations and took the statutory form of laws that permitted entry on another's lands to construct a dam, or laws limiting riparian rights to enjoin interference with the flow of streams thereby making it far less costly to obtain flowage easements for impounded water. Later statutory forays into the preference arena include statutes favoring agricultural uses and statutes favoring mining uses. Thus, through legislative intervention, the class of preferred riparian uses has grown from the traditional absolute preference for domestic ("natural") use to a variegated series of preferences reflective of state policy about the proper allocation of water resources among competing activities.[8]

Except for these preferred uses, courts initially treat all riparians equally, regardless of the amount of riparian land owned by each. Judicial reallocations of water during shortages are based only on comparative reasonableness of uses.

WHICH USES BY A NONRIPARIAN ARE PERMISSIBLE?

Nonriparians are at a grave disadvantage under the riparian system. In most riparian states water rights are not transferrable apart from riparian land, although a few states have allowed riparian rights to be severed from the land and sold independently.[2] In these minority-rule "jurisdictions" (states) the purchaser of a water right may make any reasonable use of water that his seller could have made.

Legislatures customarily give municipalities the power to acquire water rights by eminent domain, and once water rights have been condemned, municipalities may sell water to nonriparians regardless of place-of-use restrictions. *Privately owned* water companies diverting water for public sale have been given water rights under the doctrine of "prescriptive rights."[2] A prescriptive right to water arises when a nonriparian or riparian uses water to which he is not legally entitled, as if it were his own. The adverse use must be open and continuous for a period of time fixed by statute, often 15 years, in order to become a prescriptive right.[9] The true owner can terminate the adverse use by suing to stop it before the prescriptive period expires. In order to establish a prescriptive right, the riparian owner must be capable of recognizing that his rights have been invaded, and so prescriptive rights do not run upstream. One reason for the decline of prescriptive water rights is that modern state legislatures also give private water companies powers of eminent domain.[2]

EVALUATION OF RIPARIANISM

The riparian system favors flexibility over security. By and large it has worked well in the humid, water-rich eastern states, and at least one commentator believes that it is adequate to support water development in the future.[13]

However, most water diversion law experts see the pure riparian system as obsolete. The most frequent objection is that "(t)hese doctrines presume a surplus of water and are designed to resolve the occasional dispute between individual water users. Because they do not establish comprehensive regulation, these doctrines are not suitable in periods of chronic water shortages,"[12] such as those that have periodically appeared in the East since 1960.

Allocation decisions in pure riparian states are made by the courts, an institution lacking the expertise and administrative continuity to assure a predictable diversion rights system. Case-by-case judicial decisionmaking results in inconsistent and impermanent results. Any allocation may be altered by the entry of new users, changes in patterns of use, or changes in the characteristics of the watercourse.[12] This absence of definite, quantifiable diversion rights inhibits investment. It also precludes drought planning and management, because court decisions on water allocations are ad hoc and restricted to actual litigants. Comprehensive record-keeping and water supply planning are impossible in a pure riparian state. This is exacerbated by riparianism's failure to recognize the relationships between surface and groundwater.

Riparian place-of-use restrictions are anachronistic. They originated in preindustrial England, where only lands immediately adjacent to waterbodies could be the sites of viable uses. This is no longer the case.

> The arbitrariness of limiting water rights to parcels adjoining the water is, at least from a modern perspective, evident. First, there is no necessary correlation between contiguity of land and water and the ability to benefit from use of the water. Second, there is a degree of fortuity in the patterns of land ownership that result in the owner of a large single estate, part of which is riparian, but part of which lies far from the water's edge, having the entire tract considered riparian while a less fortunate landowner whose entire parcel lies close to the water, but without touching it, has no riparian rights.[8]

The watershed limitation arbitrarily discourages interbasin transfers of water, while the nontransferability of riparian rights apart from the land frustrates market solutions to water shortages.

Finally, public rights and fish and wildlife habitat are inadequately protected by the riparian system. Although "the instream flow needs of recreational riparian users are often protected,"[13] nonriparian members of the public have no standing to contest excessive diversions by riparians. Without enabling legislation, even a state might be incapable of safeguarding instream values unless it owns riparian land itself.

Thus, eastern states, beset by drought and increasing demands for water, have moved toward permit programs that add a measure of security but retain some of the flexibility of riparianism.

Chapter 3
WATER DIVERSION DOCTRINES: EASTERN PERMIT SYSTEMS

Fifteen of the twenty-five eastern riparian-rule states, and Arkansas and Iowa west of the Mississippi, have enacted administrative permit systems that allow for the management and allocation of water substantially independent of the riparian doctrine.[14] Despite superficial similarities, a number of significant differences exist between eastern and western diversion permit systems, the most important being that "the principle of priority, a critical element of western water law, is not fully recognized in the East"[15] (see Chapter 4).

Eastern diversion permit systems are highly variable from state to state, and it would be impossible in a work of this nature to thoroughly consider all the relevant state lawmaking. This chapter will present an analysis and evaluation of eastern permit systems based on categories and data developed by Sherk,[14] Davis,[12] and Ausness.[15] Examples from particular states will be given to highlight points of interest.

PERMIT AND REGISTRATION REQUIREMENTS

In all eastern states except Florida, the permit system is administered by a statewide administrative agency. Florida relies on five regional water management districts to allocate water.

Permit applications require information on the diversion location, the volume and flow rate of the diversion, and the location and nature of the water use. Once a permit has been issued, diverters are required to monitor their diversions and uses and to maintain records and make periodic reports. Using this information, the agency that administers the permit program can establish a data base for water quantity planning and management. Quantification of diversion rights and the identification of minimum stream flows are also facilitated.

Place-of-use restrictions are abolished in a number of eastern states, including Florida, Kentucky, and New Jersey. In states such as these, the agency is empowered to deny or place limitations on permits for interbasin transfers of water.

Most eastern permit statutes provide for application fees, but few require water use royalties that might encourage efficient use of water.

In all but a few eastern permit states, failure to apply for a permit by a specified date invalidates any prior claim to the use of water. Put another way, all water diversion rights that are not integrated with the permit system are conclusively presumed to have been abandoned. Several states (New York, for example), have merely enacted registration requirements without compliance deadlines or forfeiture penalties for nonregistration. "Register it or lose it" statutes are constitutional.[9]

INCLUDED WATERS AND USES

Most eastern permit statutes regulate diversions of both surface watercourses and groundwater. Wisconsin and Arkansas require permits only for diversions of surface water; South Carolina and Virginia only for groundwater. Georgia and Illinois deal with surface and groundwater in different statutes. The Indiana and North Carolina permit programs cover only "critical areas." The Delaware and Florida statutes apply to diffused surface water, but most other states expressly exempt diffused surface water. "Other typical exemptions include farm ponds, surface watercourses located on single-ownership tracts, and springs."[12]

Diversions for domestic purposes ("natural uses") are exempt from permitting in most states. Four states exempt all agricultural diversions, including for irrigation. Two states exempt diversions for public water supply. Kentucky exempts water diversions for steam power plants and oil and gas recovery.

Most states have established minimum use quantities to which the statutory requirements apply. The preferred threshold is 100,000 g.p.d. Below this threshold, applicants are either wholly or partially exempt from permit requirements.

Properly registered "grandfather rights"—diversions existing at the time of enactment—are constitutionally sacrosanct, at least up to amounts of water reasonably required for the pre-existing uses. "All such pre-existing diversions and uses are vested rights and must be granted permits."[12] Divestment of a grandfather diversion right would constitute a "taking" for which compensation would be necessary. Wisconsin limited grandfather claims to the maximum amounts of water diverted prior to the statutory enactment date. In New Jersey, the agency may restrict grandfather rights "to the extent currently exercised, subject to contract, or reasonably required for demonstrated future need."

ALLOCATION OF WATER

Only one eastern permit statute contains an explicit set of priorities for allocating water during periods of shortage, although statutory exemptions for agricultural and other diversions function as preferences. Minnesota's statute applies to all diversions. During drought periods or when conflicts exist between competing permit applications, the "preferred uses, in descending order of priority, are: (1) domestic water supply, excluding industrial and commercial uses of municipal water supply; (2) any consumptive use of less than 10,000 gallons per day; (3) agricultural irrigation and agricultural processing; (4) power production; and (5) other uses." Most other eastern state legislatures have delegated power to administrative agencies to allocate scarce water based on general policy guidelines such as "the public interest," "comparative public benefits," "reasonable beneficial uses," "fair share allocation," and the results of "state comprehensive water planning." A few eastern permit statutes set out factors to be considered by the administering agency. "This list of factors is very similar to lists recited by courts in riparian rights cases."[12]

Only Mississippi among eastern states has adopted the prior appropriation system. However, in other eastern permit states, a perfected grandfather use will supersede a postenactment use where the two conflict. Other states, either by statute or regulation, require permit applicants to show that their proposed uses would not interfere with existing uses.

FLEXIBILITY VERSUS SECURITY

Security of investment may be achieved by setting a permit term long enough so that the permitee's investment may be amortized. In this regard, seven statutes provide for fixed term permits, with maximum terms of 10 or 20 years. In three states, municipal water suppliers may obtain 50-year permits. Where state statutes are silent about permit terms, permits are either granted in perpetuity or for terms set by administrative agencies. In three states, permits with indefinite terms must be reviewed periodically and may be cancelled. "Fair compensation" must be paid in New Jersey when a permit is cancelled. The North and South Carolina statutes provide that permits may be granted for 10 years, the duration of the critical area designation, "or a sufficient time to allow for amortization of the applicant's water withdrawal or water use facilities, whichever is longest."

When a permit expires, the administrative agency may renew the permit or reallocate the water to a new use that is more consistent with the public interest.

Several eastern permit statutes contain provisions, similar to those in prior appropriation states, regarding forfeiting diversion rights for nonuse. In Maryland and Minnesota, permittees must initiate diversions within two or five years respectively. In Florida, Georgia, and Iowa, cessation of use for two or three years causes a permit to be forfeited.

Transferability of permits enhances both flexibility and security. More socially beneficial users could buy out existing ones in shortage situations. Secondly, low-priority users could recover compensation for their unrealized investment by selling their rights instead of suffering uncompensated permit cancellation or reallocation after expiration. Nevertheless, few eastern diversion permit statutes allow transfer of permit use rights to new locations or uses. In North and South Carolina, permits may be transferred with agency consent. The New Jersey agency is only authorized to consent to a permit transfer for an identical use of water.

Finally, security is enhanced because subsequent applicants must ordinarily show that their requested withdrawals will not harm existing water users.

EVALUATION OF EASTERN PERMIT SYSTEMS

None of the eastern permit statutes has been tested against a serious drought. Will they be fair and effective when difficult allocation decisions must be made?

Exemptions of particular sources and uses are antithetical to comprehensive water resources planning and management. Agriculture and public water supply exemptions, for example, discourage conservation by these highly transformational users. Grandfather rights may be constitutionally entitled to protection, but their existence will limit management options during periods of scarcity.

Preferred uses should be stipulated by legislatures. "Because basic economic and social issues are involved in answering fundamental water allocation questions, the legislature should make these allocation policy decisions"[12]. Instead, all but one eastern legislature have left these controversial decisions to be made by administrative agencies, in the political heat of a drought, with only generalities such as "public interest" for guidance. The same absence of legislative direction handicaps agencies when they consider cancellations or renewals. According to Professor Ausness, eastern permit statutes do not provide a viable mechanism for reallocating water to more productive uses.[15] Moreover, permits are not well integrated with state water resource planning.

Ausness is also critical of these systems because, in his opinion, their short permit terms, agency reallocation options, and prohibitions on voluntary transfers do not afford sufficient security to permit holders. He recommends variable-term permits such as North and South Carolina employ, compensation for permittees whose permits are cancelled or not renewed, and voluntary transfers, subject to agency approval.

There are also unanswered constitutional questions about whether postcancellation and postexpiration reallocations are "takings" for which either government or the new permittee must compensate the former permittee. No eastern cases have been found that bear on this precise issue, but the Florida Supreme Court's 1978 decision in *Village of Tequesta* v. *Jupiter Inlet Corp.*[16]—a case involving extinguishment of unregistered groundwater diversions—may be an indication of judicial responses to reallocation.

Jupiter owned property near Tequesta on which it planned to build a 120-unit condominium project. This property was located approximately 1,200 feet from Tequesta's well field. Tequesta's wells, 75–90 feet deep, pumped in excess of a million gallons of water a day from the shallow aquifer to supply Tequesta residents with water. As a result of Tequesta's withdrawals, salt water intruded into the shallow aquifer. Jupiter was forced to drill a well to the Floridian aquifer located 1200 feet below the surface. Of course, Jupiter's drilling and pumping costs were much greater than they would have been had Tequesta's pumping not caused contamination of the shallow aquifer. Jupiter sued Tequesta for taking its property without just compensation. The Florida Supreme Court decided in favor of Tequesta, holding that Jupiter had no legal right to use the water beneath its land; that is, Jupiter possessed no property that Tequesta could have taken. According to the Court, the Florida Water Resources Act, requiring diversion permits, replaced the system of common law rights that preceded it. The act provided that a common law water right would terminate unless converted into a permit water right within two years after its passage. Tequesta had perfected its water right by acquiring a permit in timely fashion, while Jupiter had lost its unexercised common law right by not filing for a permit in time. Thus, Jupiter had no water rights as against Tequesta, a permittee. Given the severe threats to public welfare posed by water shortages, reasonable reallocations are probably also constitutional without compensation.

On the positive side, eastern diversion permit statutes retain a good measure of the riparian system's flexibility while affording a permittee a quantifiable water right for a fixed or reasonable term. Courts are, to a great extent, removed from the water allocation process, and the capabilities of state administrative agencies to plan for and react to water shortages are greatly improved. Conjunctive management of surface and groundwater is expedited in most states, and obsolete place-of-use restrictions are abolished in others. Additionally, many of these statutes establish procedures for protecting public rights by maintaining minimum instream flows. In fact, some states relate allocation decisions to comprehensive state water management plans that include instream uses.

It remains to be seen whether eastern diversion permit statutes are powerful and equitable tools for water resources management or, as Meyers and Tarlock conclude, "function less to allocate water than to provide systematic information for state water planning."[17]

Chapter 4

WATER DIVERSION DOCTRINES: PRIOR APPROPRIATION

The prior appropriation (or "appropriation") doctrine is a water allocation system that developed as an adaptation to conditions in the western United States. The early settlers of the West quickly recognized that the riparian doctrine was unsuited to much of the newly-opened territory.

> The first settlers for whom water was a problem were the miners who arrived beginning about 1850, and agriculturalists who came shortly thereafter. Several realities confronted these new residents: (1) they did not own land, virtually all of which belonged to the United States government as a result of the Louisiana purchase and the Mexican cessions; (2) the United States had not yet begun the major programs, such as the Homestead Act and the federal mining laws, which would permit private land acquisition; and (3) the places in which the settlers needed to use the water were often located quite a distance away from the rivers.[8]

In true pioneer fashion, these settlers simply entered federal lands and diverted needed water. They applied the same property principle to water as they did to minerals and land: "First in time is first in right." This "customary law," or "law of the mining camps," was later ratified in federal and state statutes and western state constitutions.

However, the abandonment of riparianism in favor of the appropriation doctrine was not uniform throughout the West. Riparian theories of diversion and distribution were ignored entirely in the eight more arid states

(Colorado, Arizona, Idaho, Montana, Nevada, New Mexico, Utah, and Wyoming). These are generally referred to as "Colorado doctrine" states. In these states, the appropriation doctrine has always been recognized as the exclusive means by which the right to divert may be acquired. However, in the generally less arid states east and west of the Colorado doctrine bloc (California, Kansas, Nebraska, North Dakota, Oklahoma, Oregon, South Dakota, Texas, and Washington), the riparian doctrine was recognized in greater or lesser degree along with prior appropriation.[18] Over the years, all of these "California doctrine" states except California itself have abolished unused riparian rights and integrated existing riparian uses into their permit systems.[19] "Only California has retained a truly mixed system of riparianism and prior appropriation in which water users can initiate new uses based on either of the two regimes."[8] In California, riparian rights are generally superior to all appropriation rights. Alaska qualifies as a Colorado doctrine state; Hawaii possesses a unique system based on local customary law.

APPROPRIATIONISM AND RIPARIANISM

The basic principles of appropriationism apply in both Colorado and California doctrine states. After the principle of priority in time, the second major principle of the appropriation doctrine is that "beneficial use shall be the basis, measure, and limit to the use of water."[20] The interplay between these two principles explains the four major differences between appropriationism and riparianism.[21]

First, since appropriation rights are based on priority of beneficial use, not on ownership of riparian land, anyone can acquire an appropriative right for use at any location. Realistically, appropriative rights are limited only by the economics of applying water from a particular source for use in a particular place.

Second, an appropriative right exists for a definite amount of water and is not correlative with other diversion rights. If an appropriator can make beneficial use of all unappropriated water in a watercourse, he can lawfully appropriate the entire quantity unless the state imposes a "minimum flow" restriction. When a water shortage occurs, the burden falls on "junior appropriators" (later diverters): they are closed down as necessary in inverse order of priority, i.e., the latest allocation granted is the first to be closed. When this occurs, a senior appropriator is said to have "put a call on the river." There is no sharing the burden of shortage as in riparianism. The "senior appropriators" are never closed down unless the watercourse virtually dries up. The term "senior appropriator" refers to order of temporal priority, not location; it is possible to be either an "upstream senior" or a "downstream senior." A senior appropriator may take water needed by a junior appropriator below him, while the junior appropriator must permit the water to flow past his headgate if needed by a downstream senior.

Third, an appropriative diversion right, being separate from land ownership, is transferable apart from the land in most western states. The present rightholder is legally entitled to change—or sell or lease the power to change—the nature of use, place of use, point of diversion, or method of diversion.[20] However, state administrative agencies will not allow transfers unless senior and junior water rights in return flow are protected. This is referred to as the "no injury" rule,[22] an example of which is given by Meyers and Tarlock[17]:

> Suppose A Canal Co. obtained an appropriative right on Clear River in 1867 to divert 20 second feet of water during a growing season of 120 days. . . . Thereafter, B Canal Co., knowing these facts, obtains a right for its downstream ditch to 10 second feet. Though B does not know it, 10 second feet returns to the stream from A's diversion, and this return flow enters the stream above B's headgate. A proposes to sell his water right to Central City, which plans to move the point of diversion to an intake below B's canal. If A can move all 20 second feet, B will have no water supply when the flow of the river at A's old point of diversion is just 20 second feet. Central City could require the 20 second feet to by-pass B's headgate, thus depriving B of the return flow from the old diversion. Such a transfer is not permitted; only A's consumptive use of 10 c.f.s. may be transferred, leaving 10 second feet to substitute for the return flow.

The no injury rule can raise insurmountable obstacles to diversion rights transfers, especially in large watersheds. "One need only imagine a river with dozens or even hundreds of diversions, all producing and using return flow, to appreciate how difficult it can be to determine the impact of adding and/or subtracting any single user, and having to assure that no appropriator will be worse off after a transfer."[8] Moreover, western water-use records describe water rights in terms of rate or quantity of diversion, rather than consumption. Since only amounts consumed may be transferred, the inadequacy of records further complicates transfer proceedings.[22] Other restrictions imposed by the Bureau of Reclamation, irrigation districts, and states also impede transfers of western water rights. (For example, Wyoming and South Dakota tie irrigation rights to particular pieces of land, and Montana forbids transfers from agricultural to industrial uses.)

Finally, an appropriative water right is of indefinite duration as long as it is used in accordance with the law. However, unlike a riparian right, an appropriative right can be lost through nonuse because its legal existence was created by actual use only.[20] An appropriative right can be lost by abandonment, forfeiture, or prescription. Should a rightholder not use his right, or part of it, for a statutory period of time, and intend not to use it, his water right may be lost by abandonment in a proceeding brought by the state or a junior appropriator. Intent is the key to abandonment.[23] Forfeiture is a statutory remedy for nonuse and only requires a showing of nonuse for a specific period, usually three to five years. No intent need be proved apart from actual nonuse. Both riparian and appropriative rights may be lost by prescription. Prescription depends more on an adverse use than on nonuse by the rightholder. But prescription is as disfavored in western water law as it is in the East.[23] Abandonment, forfeiture, and prescription are issues frequently raised in transfer proceedings, further confusing that already cumbersome mechanism.

BENEFICIAL USE

Appropriative rights depend on beneficial use, but what is beneficial use? Although state statutes and court decisions identify some beneficial uses, the term is operational and must be tested in each case.[20] In the past, there has been a pronounced economic element to beneficial use; domestic, municipal, irrigation, stockwatering, power generation, mining, and milling uses have been paramount. But in recent years recreation, scenic beauty, and ecological protection have been recognized as beneficial uses. For example, in 1974 the Idaho legislature directed the Department of Parks to appropriate certain unappropriated waters of Malod Canyon in order to maintain instream flows, declaring that "the preservation of water for scenic beauty and recreation uses is a beneficial use of water." Although an actual physical diversion is generally necessary to perfect an appropriative right, the Idaho Supreme Court upheld the statute.[24] Storage of water for future uses has long been considered a beneficial use, but storage and delayed use cannot impair the rights of other appropriations. Some states have their own unique variations on the beneficial use theme.

In theory, no appropriative right attaches to water in excess of the quantity used beneficially. But waste of water is a significant problem in the West, especially where irrigation is concerned. To some extent, waste of water is endemic to the appropriation system. Extensive losses from seepage and evaporation may be caused by prohibiting an upstream junior appropriator from diverting water intended for a senior appropriator whose headgate is located many miles downstream. But the system as administered has also encouraged waste. Hutchins quotes a reclamation engineer of many years experience to the effect that "padding" and "pyramiding" water rights—creating records of rights in excess of actual beneficial use and filing and maintaining in good standing applications to appropriate more water than is required for a particular purpose—have been commonplace in appropriation states.[25] Padding and pyramiding have been tolerated, in part, because the criterion for beneficial use has been inefficient local irrigation practices. Moreover, in some western states like Arizona and Colorado, salvage water is subject to appropriation by the public; it cannot be reused by the one who conserves it. Combined with the possibility of forfeiting unused water and the requirement that downstream appropriators of irrigation return flow must be protected, the rule against use of salvage water by the conserver has induced waste. Why invest capital to save water that will benefit others?

Nevertheless, states have refined the concept of beneficial use to combat waste. One of the earliest of these attempts was the "duty of water" statutes.[20] "Duty of water" is defined as the quantity of water essential to the irrigation of any given tract of land. Many western states enacted statutes prescribing maximum quantities per acre that may be appropriated for irrigation purposes, expressed in cubic feet per second for direct use and acre-feet for storage and subsequent use. But duties of water, based on local custom, have been set at inordinately high levels. Moreover, some appropriators who did not need the maximum quantity have been awarded it anyway, while others who actually required more than the maximum have been precluded from obtaining it. Some western states are moving toward more flexible duties of water. New Mexico, for example, defines it as an amount "diverted at a rate consistent with good agricultural practices and which will result in the most effective use of available water to prevent waste."

More important in preventing waste is the concept of "reasonable beneficial use." The California constitution makes reasonable beneficial use the limit of every stream water right, and reasonableness of use is playing an increasingly important role in other appropriation states. The element of reasonableness in western water law does not connote, as it does in riparian states, a correlative and therefore uncertain right to use water. Reasonableness in the West means, in the words of a Texas statute defining beneficial use, "the use of such a quantity of water, when reasonable intelligence and reasonable diligence are exercised in its application for a lawful purpose, as is economically necessary for the use." The arbiters of reasonableness in the West are generally state administrators who possess broad discretion to grant, deny, or condition new diversion permits in the public interest.[25] Negotiations aimed at eliminating waste are becoming an integral part of the permitting process. Results may include permitting diversions of different amounts at different times of year. Rarely do courts become involved. Even when they do, a fundamental principle of administrative law is that an administrative decision will not be overturned unless it is a violation of statute, a deprivation of constitutional rights, or "arbitrary and capricious" (clearly irrational or based on an inadequate record). A court will ordinarily not substitute its judgments for those of an administrator, especially where a legislature has delegated discretionary judgments to administrators based on their presumed expertise.

"Preferred uses" are pronounced by statutes and state constitutions to be more beneficial than other uses. One effect of a water right being preferred is that in times of shortage, a preferred use may condemn a non-preferred use in order to supply water for the higher use, regardless of temporal priority of diversion. When a preference is exercised, all or part of a nonpreferred right may be taken temporarily or permanently.[23] However, compensation must be paid for taking a water right. Another function of legislatively established preferences is that they serve as criteria for the allocating agency when rival applicants are competing for the same unappropriated water for different uses. A viable system of preferences is critical in the West because there is little usable, unappropriated water remaining in most western states. As we have seen, a major shortcoming of eastern diversion permit statutes is their lack of legislatively delineated preferences. The order of preference may vary among western states, but normally the ranking is (1) domestic use; (2) agricultural use; (3) industrial and power use; (4) fish, wildlife, and recreation uses. There is an obvious bias in favor of domestic and economic uses as against ecological and recreational uses. This emphasizes the need for establishing minimum instream flows.

WATER CODES

Colorado is the only western state where the pure appropriation doctrine, under which a person obtains a water right simply by using water, still exists.[26] All other western states have adopted "water codes" (statutes) that govern: (1) approving appropriations of water, (2) determining or adjudicating water rights among conflicting claimants, and (3) administering water rights and distributing water.[25]

The first step in obtaining an appropriative water right is to file an application with the state water official, normally the state engineer. In several states there are steps that can be taken in advance of filing the application, e.g., "notice of intention" in New Mexico, designed to give a prospective applicant time to make investigations, while holding priority of filing, before finalizing the application. If a supply of unappropriated water is available, the state engineer publishes a public notice to provide an opportunity for protestors to be heard. Protest proceedings may be especially complicated in the California doctrine states of California and Texas, where riparian rights are still of major significance.[20] Where the proposed use is in conflict with prior diversion rights or antithetical to the public interest, the state engineer may reject the application. Wasteful diversions or uses could be denied at this stage.

If protests are resolved in the applicant's favor, and the use is reasonable and beneficial, the applicant will then receive a permit to proceed with the project. The permit specifies the approved project and stipulates that construction must proceed with reasonable diligence and be completed within a reasonable time, generally five years unless circumstances justify an extension. In most western states a permittee may also be granted the power of eminent domain to condemn private land for his ditch right-of-way.[27] The appropriation is incomplete until water has actually been diverted and applied to a beneficial use. There are exceptions to the "actual diversion" requirement for instream uses in almost all western states. When the permitted diversion is finished, the water right, in the form of a document called a "license," "certificate of appropriation," or "water right certificate," is issued. This document "relates back," for purposes of priority, to the date of application or notice of intention. Once perfected, an appropriative diversion right continues until transferred, condemned, or abandoned. Municipal water supply planning is facilitated by allowing municipalities to obtain "conditional water rights." By this means, municipalities may reserve increasingly scarce water for future expansion even though they presently do not need additional water or are unprepared to construct diversion works.[26]

Statutes of all 17 western states contain some kind of special procedure for adjudicating water rights.[25] Adjudication is the formal process of settling disputes over, describing, and recording diversion rights in a particular area or basin. An adjudication results in a decree specifying the amounts and priorities of diversions on a watercourse. This process is not uniform among the western states, but three general models can be identified.[20]

1. Private litigation between two or more water rights claimants or users. No participation by state officials is contemplated.
2. Proceedings under statutes designed to permit or encourage official participation in private actions. The purpose of these statutes is to adjudicate as many rights as possible in the fewest lawsuits. A court may be permitted to convert a private suit among a few individuals into a complete adjudication of a watercourse, using the expertise of state officials.
3. Proceedings under statutes authorizing either state action or petition of water rightholders to adjudicate an entire watercourse.

Where the third model has been adopted, a state administrative agency, rather than a court, performs the adjudication. Colorado calls its system of water judges, water clerks, special engineers, and referees "judicial," but it functions as an administrative agency.

Administration of the appropriation system is improved where a watercourse has been adjudicated. But "(e)ven if it worked perfectly, which it does not, the general adjudication system would provide an incomplete record of water rights. . . ."[8] First, not all streams have been fully adjudicated, and some adjudications are out of date. Second, many older adjudications reflect inflated diversion claims. Third, some adjudicated diversion rights have been abandoned or forfeited. Finally, several classes of water rights, such as federal reserved rights, groundwater rights, and prescriptive rights, were not traditionally included in general adjudications. "Pueblo rights" are another type of unadjudicated water diversion rights. In California and New Mexico, cities that had been pueblos under Spanish law "have a water right [for municipal water supply] superior to all other riparian and appropriative rights, which right can be exercised even though others have previously been using the water."[8] Unused riparian rights would not be covered by California adjudications.

Once a watercourse has been adjudicated, how are the complicated determinations to be enforced? Statutory procedures for administration of water rights are in effect throughout the West.[25] In some states supervision by administrative officials goes beyond adjudicated rights and extends to unadjudicated rights evidenced by permits, licenses, or certificates of appropriation. State water officials are also called upon to enforce rotation plans imposed by contract or court decree.[27] Rotation is a means of enhancing the efficiency of irrigation by giving an irrigator a large quantity of water over a short period of time rather than a lesser amount, to which he holds a right, for a longer time. Under a rotation schedule, each user is entitled to divert the entire flow of the stream for a period computed by the ratio which his appropriative right bears to all rights involved in the schedule. Although an apparent departure from strict appropriative allocation, rotation is approved by legislation in many western states. To the same effect, modern appropriation permits and certificates specify the times of year when diversions may be made.

Some western states create water-supervision divisions or districts and provide for enforcement agencies.[25] Others authorize the state engineer to create districts where necessary. Methods of selecting watermasters, commissioners, or patrolmen, as they are variously termed, and of allocating costs of supervision, vary among states. Watermasters regulate diversion from a watercourse, opening and closing headgates according to detailed schedules of priority. They also have power to make arrests for tampering with legally controlled headgates.

CLASSES OF WATER

Early constitutional and statutory provisions in the West declared that all unappropriated flows of surface water in watercourses were subject to appropriation.[19] A strict interpretation of this language explains why some western states do not permit percolating groundwater to be appropriated. Diffused surface water is not available for appropriation before it reaches a watercourse. Diversion rights to spring water have been as troublesome in the West as in the East, and a number of western states have enacted statutes dealing with spring water. In California, where riparian rights are valuable, the law excludes from appropriation water which is "reasonably needed for useful and beneficial purposes on lands riparian thereto."

It has been pointed out that where irrigation return flows are concerned, junior appropriators have legal rights in the continuation of stream conditions that existed at the time of their respective appropriations; they may later resist all proposed transfers by upstream irrigators that would injure them. Once again, in many western states salvage water is immediately subject to appropriation, and the salvager cannot reuse it. Wastewater and return flows from foreign and developed water are treated quite differently. The "generator" of these waters may use, reuse, use for a different purpose, or dispose of the waters until he abandons them without intent to recapture. Once these waters pass from the generator's land or reach a natural watercourse they are presumed to be abandoned and become subject to appropriation. States differ as to whether these flows are governed by the existing priority system, constitute "new water" belonging to the first appropriator, or must be allocated among all users. However, in contrast to irrigation return flows, downstream appropriators have no rights to the continuation of waste discharges or return flows from foreign or developed water, and the generator can discontinue these flows at any time. Western states handle the appropriation of storage water in different ways.[27] One method makes no distinction between direct flow and storage rights, integrating them as steps in the acquisition of a single appropriative right. In these states the procedures of diverting, impounding, distributing, and applying the water to beneficial use are simply phases of one complete administrative procedure. Another method deals with diversion for storage and distribution of stored water as separate segments of an overall plan and provides separate but complementary procedures in the form of "primary" and "secondary" permits. The final certificate of appropriation, however, applies to the entire project. In a few states, entirely different appropriations are involved. Whatever the administrative system for appropriating storage water, it is true throughout the West that priorities on a stream system are determined on the basis of temporal priority of appropriation, regardless of whether appropriations are direct flow, storage, or both. Nor is there a preference for on-channel or off-channel storage. A number of western states also make provision for storing water in natural lakes and underground, as long as the rights of prior appropriators of water from these sources are protected.

EVALUATION OF APPROPRIATIONISM

According to Professor P. N. Davis, "the principal defect of prior appropriation is the result of its primary virtue. The absolute security of a water right held by the senior appropriators makes it very difficult to establish new uses."[12] This inherent inflexibility of the appropriation system is exacerbated by the frequently insurmountable obstacles placed in the way of diversion rights transfers. Preferred use statutes are a mitigating factor, but even they are rigid in contrast to the flexibility and adaptability of the free market. Appropriationism's rigidity is illustrated by its clumsy responses to modern demands for agricultural water conservation and maintenance of instream flows. Waste is a pervasive and continuing problem in the West, and the legal system has been slow to develop doctrines that might counteract it. The fledgling western water conservation movement is analyzed in Chapter 23. Various strategies for providing adequate instream flows are discussed in Chapter 17.

One of the deficiencies of appropriation law has been its tendency to pigeonhole water uses into distinct and inviolable classes. This approach confuses western lawyers and judges, not to mention water resource managers. There is much harm and little good in distinctions such as those between irrigation return flow and irrigation seepage, and between salvage water and wastewater. Nevertheless, the appropriation system has substantially fulfilled its goal—to encourage the maximum, economically beneficial use of water and promote the economic development of the West.

Chapter 5

GROUNDWATER DOCTRINES AND PROBLEMS

Because the hydrologic relationship between groundwater and surface water was not well understood, groundwater diversion doctrines evolved independent of surface water doctrines. The salutary trend toward conjunctive management of ground and surface water supplies is encouraging an integration of legal approaches. One example is the eastern diversion permit statutes, most of which apply to diversion of both ground and surface water. Other instances will be discussed in this chapter.

However, there are fundamental differences between managing groundwater and surface water resources. "Unlike surface streams, there is no short-term natural limit to the amount of a resource that can be extracted over time. Thus groundwater can be pumped beyond the rate of annual recharge."[28] Determining the limitations on groundwater overdrafting, or "mining," is a significant policy issue in groundwater management.

Second, comprehensive management of quantity and quality is even more important with groundwater than with surface water. Pollution in a surface watercourse tends to mix with, and be assimilated by, the total volume of water in that watercourse. In contrast, groundwater pollution constitutes a slow-moving "slug," because infiltration of rainwater is limited. Consequently, pumping from a polluted aquifer may pose more profound public health risks than diversion from a surface supply. Moreover, mining an aquifer may cause the intrusion of new pollutants, such as salt water, or influence the direction and speed of the pollution plume. Finally, once the pollution is abated, it takes much longer for an aquifer to be decontaminated than a surface watercourse. For all these reasons, there should be a cohesive system of legal processes and rules applicable to all aspects of groundwater use. This is not presently the case, and the organization of this book reflects the current situation. Groundwater diversion law is analyzed in this chapter, while groundwater pollution control law is taken up in Chapter 37. That chapter contains recommendations for integrating these unfortunately separate areas of law.

Six categories of legal doctrine govern conflicts in different states among competing diverters of groundwater: (1) the rule of absolute ownership; (2) the reasonable use rule; (3) the Restatement of Torts rule; (4) the correlative rights rule; (5) prior appropriation; and (6) statewide or critical areas management systems. In eastern permit states, administrative agencies will be guided by that state's common law diversion principles[10] and political considerations in making groundwater allocation decisions.

ABSOLUTE OWNERSHIP

This rule creates an absolute right in each landowner to pump groundwater under his land for use at any location without bearing any responsibility to neighboring owners, except for "spite pumping." Where there is competition for diminishing groundwater reserves, "the biggest pump wins." The doctrine of absolute ownership is based on an outdated assumption—the movement of groundwater is unknowable and unpredictable; thus a landowner, who is legally entitled to everything below the surface of his property, may use this property as he wishes because the law does not impose responsibility for unforeseeable consequences. Nonmalicious waste is not prohibited under the absolute ownership rule. Absolute ownership of groundwater has been rejected in most states, but still prevails in Indiana and Texas.[28]

REASONABLE USE

Confusingly, the reasonable use groundwater rule is both like and unlike the reasonable use surface water rule. Under the rubric of reasonable use, courts deciding surface water disputes apply a theory of comparative reasonableness under which a riparian may divert water until he interferes with the reasonable use of another riparian, and when that occurs, the previously reasonable use may be curtailed. In most eastern states, diversion off riparian land and outside the watershed are unreasonable per se. Courts adjudicating groundwater cases also talk about comparative reasonableness, but in practice reasonable use of groundwater is the same as absolute owner-ship with two exceptions: (1) waste is prohibited, and (2) water must be used on land overlying an aquifer unless it can be transported without injuring other overlying owners. A groundwater pumper in a reasonable use state may divert all the water he can pump as long as it is used on over-lying land without waste. There are no reciprocal obligations among ground-water pumpers under the reasonable use rule of groundwater as opposed to the reasonable use rule of surface water. On the other hand, both reason-able use rules contain place-of-use restrictions. About a dozen states, all of them eastern states except for Arizona, adhere to the reasonable use rule in groundwater allocation situations. Most of these states also subscribe to the reasonable use surface water rule, different as it is from the reasonable use groundwater rule.

RESTATEMENT RULE

"Restatements" of legal fields are formulations by legal scholars of what the law should be. They are influential in areas, such as groundwater rights, where the law is in flux.

Section 858 of the *Restatement of Torts (Second Edition)* articulates the rule that a well owner may withdraw water for use on his own or nonover-lying land unless other users are unreasonably affected. This section sets out three criteria for determining unreasonable interferences with a neigh-bor's use of groundwater: (1) causing unreasonable harm by lowering the water table or reducing artesian pressure, (2) exceeding the owner's reason-able share of the total annual supply, or (3) having a direct and substantial effect on surface supplies and causing unreasonable harm to surface water users. These criteria entail the same tests that have evolved for evaluating the reasonableness of surface water diversions. In effect, the Restatement rule is the application to groundwater of the reasonable use of surface water rule without its place-of-use restrictions.

The Restatement rule is preferable to either the absolute ownership or reasonable use groundwater rules because it (1) promotes conjunctive management, especially in reasonable use surface water states, (2) protects aquifers from mining, and (3) protects minimum flows in watercourses.[29] Section 858 has been explicitly adopted in Michigan, Ohio, and Wisconsin. Something quite similar also appears to be the law in Arkansas, Florida, Nebraska, New Jersey, and Missouri.[12,28]

CORRELATIVE RIGHTS

The correlative rights rule is confined to California, which has followed a mixture of riparian and appropriation surface water law. In adjudicating the rights of groundwater users, the California Supreme Court determined that in times of shortage: (1) overlying owners are entitled to no more than their "fair and just proportion" for on-site uses; (2) as between transporters out of the basin, first in time is first in right; and (3) overlying users have priority over transporters.[30] The "fair and just proportion" of an overlying owner has traditionally been "a proportion based on the ratio of the landowner's acreage to the total acreage overlying the aquifer."[10] In a 1949 California case, *Pasadena* v. *Alhambra*,[31] the California Supreme Court ascertained "safe annual yield" of an aquifer and apportioned it among all users—overlying users and transporters alike—in proportion to their uses for five preceding years. The Court has since restricted the *Pasadena* rule to situations where municipalities are not involved.

PRIOR APPROPRIATION

Although many western states have ostensibly applied the prior appropriation doctrine to groundwater, their systems are "prior appropriation in name only."[28] The commentators agree that prior appropriation is inconsistent with the nature of the resource.[10,22,28] Some western states soften the appropriation doctrine's protection of senior uses. For example, a Montana statute declares that "priority of appropriation does not include the right to prevent changes by later appropriators in the condition of water occurrence. . . . or the lowering of a water table, artesian pressure or water level, if the prior appropriator can reasonably exercise his water right under the changed conditions." Other states have adopted some form of "management doctrine," which involves "a system managed by a government official or commission which has considerable flexibility to regulate groundwater withdrawals to achieve objectives suitable to the particular aquifer."[22]

MANAGEMENT SYSTEMS

In general, the most sophisticated groundwater management systems have been developed in the West, where groundwater mining is common and conjunctive use of groundwater and surface water (imported and domestic) is frequently necessary. But eastern states such as Florida and New Jersey have also instituted management systems to cope with groundwater shortages. In New Jersey, 50% reductions in pumping from an overdrafted aquifer have been ordered while a new reservoir is being constructed. When it comes on line, nearby groundwater pumpers must switch over to the surface water supply, but their hookup and user charges will be partially subsidized by groundwater users located further from the reservoir. Florida's system will be discussed in Chapter 41.

The western management systems differ radically from one another: Arizona, Colorado, and New Mexico rely on different forms of state management; Kansas, Nebraska, and Texas utilize varieties of local management; California has a unique system administered by courts and water districts.[28] Seven other western states have modified the appropriation system by statutes that authorize setting reasonable pumping levels for all or parts of the state, but these statutes adopt different methods of allocating groundwater.[10]

Arizona's 1980 Groundwater Management Act is extraordinarily complex and demanding. Professor Getches summarizes it as follows:

> It provides for establishing "active management areas" [AMAs] that cover an area encompassing eighty percent of the state's population. These areas may be designated if overdraft exists, withdrawals threaten to create subsidence, or groundwater quality is threatened by saltwater intrusion. The management goal for critical areas is to achieve safe yield (withdrawals not in excess of recharge) within forty-five years. To reach this goal, the state director of water resources is required to formulate a management plan including: mandatory conservation by "reasonable reductions" in per capita use . . . , pump taxes with revenues earmarked for expenses of administration and augmentation plans; retirement of irrigated lands; and a requirement that new subdivisions demonstrate adequate water supply and quality for 100 years (or have a contract for Central Arizona Project water)[10].

Outside of the AMAs and other areas established by the Act, the reasonable use doctrine governs groundwater diversions. An explicit goal of the Act "is to shift groundwater from agricultural to municipal and industrial uses."[28]

Colorado is in the vanguard of integrating ground and surface water rights. Its management system depends on the unique distinction between "tributary" and "nontributary" groundwater. By statutory definition, nontributary groundwater is that which will not, within a period of 100 years, deplete the flow of a stream at an annual rate greater than one-tenth of 1 percent of the annual rate of withdrawal from the well being pumped.[26] Both surface water and tributary groundwater are covered by Colorado's prior appropriation system. But this is prior appropriation with a difference!

> Recognizing that integrated management of groundwater and surface water requires detailed judgments by administrators, the Colorado Supreme Court has acknowledged broad rulemaking authority in the state engineer. The engineer's rules should implement a policy of maximum utilization of water by considering all relevant factors, including the efficiency and expense of using a well and a balancing of the environmental effects of surface and well diversions. A senior user does not have a right to curtail junior pumping unless the engineer finds that the senior is using a reasonable means of diversion, even to the extent of being forced to change from a surface diversion to a well. But costs imposed on senior users by such requirements may be assessed against junior appropriators.[32]

Colorado also applies a modified appropriation system to diversions of nontributary groundwater in designated basins. The Colorado statute requires denial of a diversion permit only if unreasonable harm to senior rights or unreasonable waste would result.[10] Colorado's Groundwater Commission has ruled that an unreasonable groundwater diversion is one where "the rate of pumping in a three-mile radius of the proposed well would result in a 40 percent depletion of the available groundwater in the area in less than twenty-five years."[32] Unlike Arizona and Idaho, Colorado allows groundwater mining at a limited rate. There is no systematic regulation of nontributary groundwater outside of designated basins, except that these aquifers may not be depleted within 100 years.

The New Mexico statewide management system emphasizes basin-wide adjudications to define duties of water, integration of ground and surface water rights, and toleration of limited mining.[28] Prior appropriation is applied, but senior users have no absolute right to an undiminished pump lift.[10] A senior appropriator is entitled to "follow the source"—change a surface diversion to a well near the stream when the stream dries up because of junior appropriator pumping near its banks.[32] The State Engineer may issue regulations to limit groundwater mining, and the New Mexico Supreme Court has upheld a rule prohibiting a landowner from withdrawing more than two-thirds of the groundwater beneath his land over a 40-year period. Only Oklahoma allows more rapid depletion—100 percent within 20 years.

Nebraska regulates groundwater pumping through its Natural Resources Districts.[28] The districts may recommend control areas for designation by the State Department of Water Resources. Once an area has been designated, the district may impose well spacing restrictions, pumping rotations, transfer restrictions, and drilling moratoria. The district may also set its own timetable for depletion. Groundwater and surface water rights are not integrated in Nebraska.

Kansas law authorizes local residents to form Groundwater Management Districts.[10] District management plans must be approved by the State Department of Water Resources. Each district applies its own standards for preventing well interferences and depletion.

In Texas, where the absolute ownership doctrine prevails, local districts do not possess regulatory powers but stress education and technical assistance, especially with regard to conservation.

Local districts also allocate groundwater in certain parts of California, but, unlike Kansas and Nebraska, California's management districts also perform major importation, distribution, and replenishment functions. These districts will be examined more thoroughly in Chapter 9. Current California groundwater law is an elaborate mixture of correlative rights, appropriative rights, prescriptive rights, and pueblo rights. "In addition to these rights, it is the continuing jurisdiction of courts, actions of court-appointed watermasters, practices of local management districts, and economics of pumping that govern the conduct of users in adjudicated basins."[33] California's water budget is based on the conjunctive management of groundwater and imported surface water, which is often stored in aquifers. Consequently, the California legislature and courts have recognized the right of a public agency to import and store water without obligation to overlying landowners, the right to protect the stored water against use by others, and the right to recapture stored water.[10] California has also been a pioneer in the utilization of water pricing to facilitate conjunctive management.

As the preceding examples indicate, groundwater management is an extraordinarily volatile and variegated field. Professor Gould predicts that

> (t)he future will continue to see a great deal of experimentation in groundwater laws. Various management models, some providing for local control and some providing for statewide control, will be tried. Similarly, different techniques for determining the appropriate rate of withdrawal or for determining the proper point at which to protect pumping or pressure levels will be used. Nevertheless, the management model in one form or another, will be utilized in most states with high levels of groundwater use.[22]

It is also conceivable that the federal government will play a greater role in groundwater allocation because of the inefficiencies and inequities inherent in our current system of state-by-state allocation.

Chapter 6

FEDERAL AND FEDERAL-STATE DIVERSION LAW

Federal law is the supreme law of the land. States and other subfederal entities are empowered to make law, but only where the results are consistent with federal law. If a state enacts a statute that conflicts with a federal statute or infringes upon a federal power (e.g., the power to regulate interstate commerce), the state law may be "preempted"—struck down by a federal court. Federal preemption is rare in water diversion law because the federal government has traditionally deferred to state lawmaking when it comes to allocating rights to divert water, even on federally-owned land. Nevertheless, several aspects of diversion law are governed by federal law or a combination of federal and state law. These areas are discussed in this and the following chapter, as well as in Part II. Because they are anomalies, occurrences of federal preemption may cause profound disruption to the fundamentally state-administered water allocation system. On the other hand, federal intervention may be seen as an antidote to the inefficiencies and inequities of the current fragmented system.

FEDERAL RESERVED RIGHTS

The U.S. Supreme Court has decided that where the United States has withdrawn public land from settlement or disposal and reserved it for a specific public purpose such as an Indian reservation, national park, national forest, or the like, the federal government is presumed to have reserved enough unappropriated water to accomplish the purposes of the reservation. This implied federal reservation of water rights applies both to surface water and groundwater. It takes effect on the date of the land reservation and exempts reserved waters from appropriation under state law. It is unnecessary for the federal government to perfect its reserved right by applying for a state diversion permit.

For example, Devil's Hole, a deep cavern on federal land in Nevada containing an underground pool inhabited by a unique species of desert fish, was reserved as a national monument by presidential proclamation in 1952. In 1968 nearby ranchers, holding state appropriation permits, began pumping groundwater from the same source as Devil's Hole, reducing the water level in the underground pool and endangering its fish. The U.S. Supreme Court granted the federal government an injunction limiting pumping by the ranchers because in 1952 sufficient water had implicitly been reserved to maintain the level of the underground pool.[34]

Federal reserved rights for other than Indian reservations have not given rise to much litigation. In the first place, the United States Supreme Court has interpreted the doctrine narrowly where Indians are not concerned. For example, in *United States* v. *New Mexico*,[35] which involved reserved rights for a national forest established in 1899, the Court denied federal claims of instream flow reserved rights for wildlife, recreation, aesthetics, and stockwatering, because the only explicit purposes of the Forest Service Organic Act of 1897 were insuring a timber supply and protecting watersheds. In the second place, most wilderness areas and national forests are located in the upper sections of watersheds where there are no upstream private water users to be harmed by maintaining instream flows.

Indian reserved water rights, however, is one of the most controversial policy issues in the West. These rights are generally "inchoate"—unexercised and unquantified. Moreover, Indians are potentially entitled to enormous and unpredictable quantities of water. As Professor Getches points out,

> (t)he problem of varying water needs is especially great in the case of Indian reservations. Because the purposes are usually so broad—ensuring a permanent homeland and livelihood for the tribe—water requirements may change greatly as uses for these purposes change. Tribal uses may change from a fishery demanding constant minimum flows but no consumption, to agriculture requiring tremendous seasonal diversions with some return flows, to industrial uses consuming significant quantities of water on a constant basis.[10]

The United States Supreme Court has held that five Indian reservations on the Colorado River were established to facilitate agriculture, and thus the tribes are entitled to enough water to irrigate all the "practicably irrigable acreage." This standard generates comparatively modest and quantifiable claims. But most Indian reservations were established for more ambiguous purposes, such as "advancing the civilization of the Indians."[10]

The inchoate nature of Indian water rights is antithetical to the security and predictability offered by the prior appropriation system. Thus, western water users clamor for quantification of Indian rights. The most popular method of quantification is by adjudication. Under the so-called McCarran Amendment,[36] the federal government has consented to its reserved rights— including Indian reservation reserved rights—being adjudicated in state judicial (but not administrative) proceedings that apply to entire watersheds. Montana and Utah have relied on negotiated agreements with Indian tribes to achieve quantification. Before these agreements become enforceable they must be approved by at least the Secretary of the Interior, and perhaps by Congress itself. Congress could also quantify Indian reserved rights by legislation. Once reserved rights for Indian reservations have been quantified, they may be applied to any water uses chosen by the tribes. Regulating water use on reservations is a tribal prerogative; states have no power over water use, derived from federal reserved rights, on Indian lands.

Quantification is the major problem posed by federal reserved water rights for Indian reservations, but it is not the only one. The scope of these rights also remains to be determined. Do they apply to groundwater and surface watercourses not adjacent to reservations? In addition, may these rights be transferred to non-Indians, and if so, under what circumstances?

Professor Gould concludes that "(m)uch remains unknown about Indian water rights. Because they are a creation of the United State Supreme Court, no issue can be said to be definitively decided until the Court has spoken. Thus, Indian water rights are certain to consume much effort for some time to come."[22]

FEDERAL REGULATORY WATER RIGHTS

This is a phrase coined by Professor Tarlock,[37] that refers to water rights created by the operation of federal law, especially environmental law. In the case of *Riverside Irrigation District* v. *Andrews*,[38] Riverside had acquired a conditional water right under Colorado law to build a reservoir on a tributary of the South Platte River. However, the U.S. Army Corps of Engineers denied Riverside a dredge and fill permit on the ground that the reservoir would diminish flows 250 to 300 miles downstream in Nebraska, and thus have an adverse impact on the whooping crane, a species protected under the Federal Endangered Species Act. A federal district court upheld the Corps permit denial, citing obligations under the Clean Water Act and the Endangered Species Act. (These statutes will be examined more closely in Parts II and III.) The *Riverside* decision effectively gave the United States a right to minimum instream flows for protection of whooping crane habitat.

INTERSTATE WATERWAYS

Under our federal system of government, each state exercises "jurisdiction" (legal control) within its borders but not outside them. Every state administers a separate system of diversion rights governing the use of water within its boundaries. But state boundary lines have been set with only partial regard to hydrology. "Much of the water used in the United States comes from sources shared by more than one State"[39]. Whenever this situation occurs, a potential interstate problem exists. For example, what are the rights of a senior appropriator in one state against an appropriator—junior in time to him—located in an upstream state? Can an upstream state divert water for municipal supply without regard for the needs of downstream states? What are the rights of neighboring states to waters of a river or lake that forms the common boundary between them?

In the late nineteenth and early twentieth centuries, interstate water rights disputes were settled by state and federal courts in the context of lawsuits between individual diverters in different states.[39] This approach was too cumbersome, both legally and administratively, to survive into an era of increasing demands on water and increasing state involvement in water resources projects. Today, the rights of individual claimants are subsumed under the rights of their states; the contending parties are states, and the results bind all citizens of those states.[40] The three methods of resolving interstate water rights controversies are: (1) litigation between states in the U.S. Supreme Court; (2) interstate compacts; and (3) congressional allocation.

Equitable Apportionment

The U.S. Supreme Court ordinarily possesses only "appellate jurisdiction": it can only decide cases on appeal from lower federal courts; but where states sue one another, the Supreme Court has "original jurisdiction." Thus, the Supreme Court is the forum for judicial settlement of disputes between and among states over the apportionment of the waters of interstate watercourses. In this capacity the Supreme Court has refused to impose doctrinaire water law rules, but instead has borrowed a principle of international water law called "equitable apportionment."[41] The Supreme Court will apportion the waters of an interstate waterway as fairly as possible under the circumstances.

Equitable apportionment is not an entirely ad hoc process of balancing the equities in particular cases. When appropriation states sue each other, equitable apportionment begins with priority of appropriation without regard to state lines. But the Supreme Court will not go as far as to fix the respective priorities of the various individual users on an interstate waterway. It will award a "mass allocation" to each state, allowing the state to allot priorities to its share. The Supreme Court has also deviated from strict temporal priority in protecting a viable economy based on junior appropriations in one state against under-utilized senior appropriations in another state. Prior appropriation is only a guide to equitable apportionment among appropriation states. It can be varied by considering facts such as climate, nature of uses, return flows, availability of storage, degree of waste and instream losses, and comparative damage to different states. Similarly, between riparian states the Supreme Court will disregard riparian law where necessary in order to make an equitable apportionment.

Professor Tarlock has enumerated five broad principles that emerge from the Supreme Court's equitable apportionment decisions:[42]

1. In appropriation states, the doctrine of prior appropriation will be presumptively applied across state lines in small river basins.
2. The doctrine of prior appropriation will also be presumptively applied in large river basins, but the presumption is weaker on large compared to small river basins. . . .
3. In riparian states, the common law of riparian rights will be presumptively applied on both large and small river basins. As with the doctrine of prior appropriation, the court will temper the common law. . . .
4. In both prior appropriation and riparian jurisdictions, the Court retains the power to displace existing uses but this power will be exercised sparingly . . .
5. State planning to conserve existing supplies will assume a larger role in state efforts to avoid sharing duties or to impose sharing duties on other states.

The Supreme Court's attitude toward waste and conservation as reflected in its equitable apportionment decisions will be discussed in Chapter 23.

Interstate Compacts

The U.S. Supreme Court has equitably apportioned the waters of some important interstate rivers (Laramie, North Platte, and Delaware), but it has discouraged many more lawsuits for apportionment between states. The Court feels uncomfortable making legislative-type judgments based on a concept as vague as equitable apportionment. Moreover, the High Court lacks the technical resources to cope with the complicated hydrologic, economic, and sociological questions involved. There is a far better way to settle interstate diversion rights conflicts—interstate compacts.

Article I, Section 10, Clause 3 of the U.S. Constitution provides that "No State shall . . . without the consent of Congress . . . enter into any agreement or compact with another State or with a foreign power." Early in our nation's history, the interstate compact with congressional consent was used to settle disputes over navigation, boundaries, and fishing rights.[43] Compacts allocating the consumptive use of interstate waters and providing for their management, including water quality, are a twentieth century phenomenon.[44] Over 30 interstate compacts dealing with various water resources problems have followed the Colorado River Compact of 1922. These interstate compacts can be categorized as (1) water allocation compacts; (2) pollution control compacts; (3) planning and flood control compacts; and (4) multipurpose regulatory compacts. Only (1) and (4) are of interest to us in this chapter.

Most water allocation compacts are found in the West. All contiguous western states have entered into at least one interstate stream agreement.[40] The earliest of these simply allocated the waters of interstate river systems among the signatory states. More recent compacts both allocate water and create independent commissions to plan and monitor.[44] Some few, like the Upper Colorado River Basin Compact, give the compact commission the power to curtail use in times of shortage. Multipurpose regulatory compacts, e.g., the Delaware and Susquehanna River compacts, are rare. They grant broad powers to their compact commissions in all aspects of water resources management, including the authority to allocate water among the states, to regulate withdrawals of water, and to construct projects. Interstate water resources compacts vary widely with respect to compact purposes, membership and voting provisions, kinds and scope of powers conferred, funding support, and duration. A number of compacts provide for nonvoting federal representation on compact commissions, while some, the so-called federal-interstate compacts, give the federal representative equal status with the state members. Like equitable apportionment by Supreme Court decree, apportionment by interstate compact supersedes private diversion rights in the signatory states. As on the Delaware, an interstate compact can grow out of an equitable apportionment decree.

Congressional Apportionment

This has occurred only once, with Congress apportioning the Colorado River among the three states of the Lower Colorado River Basin—Arizona, California, and Nevada—after an interstate compact had divided the river into an upper and lower basin. Optimum use of Colorado River waters is beset by legal problems, most of which are discussed by Meyers and Tarlock in their case study of the issue.[17]

It is important to notice that these three methods of resolving interstate diversion rights disputes—judicial equitable apportionment, interstate compact, and congressional apportionment—have been applied almost exclusively to surface water conflicts. Only a few interstate compacts deal with groundwater at all, and these only in peripheral ways. With increased dependence on groundwater and heightened danger of groundwater over-drafts and pollution, the courts, states, and Congress will be increasingly called upon to reconcile interstate claims to groundwater resources.

INTERNATIONAL WATERWAYS

International water law governing relationships between the United States and its contiguous neighbors, Canada and Mexico, consists of treaties and the "customary international law" used to interpret these treaties.[45] The federal government has the constitutional power to enter into treaties with foreign nations; the states cannot negotiate treaties.[46] Treaties into which the United States enters with other nations take precedence over state law where the two conflict. Thus, where treaties apportion the waters of inter-national and transboundary streams, states are limited in their abilities to create diversion rights. Only uses fitting within the national share of water, and within the hierarchy of uses if set by treaty, may be effectively estab-lished by states.

The power of states to create diversion rights is primarily limited, with regard to Canada, by the Boundary Waters Treaty of 1909 and the Colum-bia River Treaty of 1961, and with regard to Mexico, by the Rio Grande Irrigation Convention of 1906 and the Rio Grande, Colorado, and Tijuana Treaty of 1944.[46] These treaties were negotiated under the principle of equita-ble apportionment of international drainage basins—the cornerstone of international water law. Important institutions established by these treaties are the International Joint Commission, which regulates the use of U.S.-Canadian boundary waters, and the International Boundary and Water Commission, which supervises the allocation of Rio Grande and Colorado River water between the United States and Mexico.

Chapter 7

INTERBASIN AND INTERSTATE TRANSFERS

This chapter examines two legal problems that affect the allocation of diversion rights in the United States: (1) the legality of major interbasin transfers; and (2) the constitutionality of state "antiexportation" statutes.

MAJOR INTERBASIN TRANSFERS

The large interbasin transfers that have already been constructed in the United States are almost exclusively intrastate (e.g., Colorado River Aqueduct and New York city diversions from the Delaware). This reflects the profound impact that political boundaries have had on water resources planning and development. There are, of course, legal and political problems with large intrastate diversions. For example, the "area of origin" may demand protective legislation, state laws may inhibit financing, and a state legislature or administrative agency may prohibit a particular interbasin transfer. However, these difficulties are relatively insignificant compared to those confronting proposed interstate diversions of any magnitude.

Professor Johnson defines "major interbasin transfers" as transfers that would carry water over one or more state lines for use in a state that either (1) lies entirely outside the basin of origin, or (2) lies partly within the basin of origin but which would import substantially more water than it contributes to the basin of origin.[47] This definition covers, for example, the following proposed diversions: (1) from the Columbia River Basin to the Southwest for use in California and Arizona; (2) from the Missouri River Basin for use in Kansas, Colorado, Oklahoma, Texas, and New Mexico; (3) from the Mississippi Basin for use in west Texas and New Mexico; (4) from the Hudson River to northern New Jersey; and (5) from the Great Lakes to the lower Midwest and Southwest. An out-of-basin state probably lacks standing to sue in the U.S. Supreme Court for apportionment of the waters of a river lying entirely beyond its borders; thus, major interstate diversions would have to be accomplished by interstate compact or congressional allocation.[47]

As a political matter, neither Congress nor the states would act affirmatively unless the area of origin consented to the transfer. Some area-of-origin protection statutes that have worked within states might serve as models for interregional approaches. One California area-of-origin statute prohibits the release of any state-appropriated water "necessary for the development of" a county of origin. Another statute gives the watershed of origin a prior right to "all the water reasonably required to adequately supply the beneficial needs of the watershed." This is known as a "right to recapture." Colorado has a statute requiring contemporaneous construction of compensatory storage dams in the area of origin so that it will be no worse off because of the diversion. Texas prohibits diversions needed to supply the reasonable future needs of the basin of origin for the next 50 years. An Oklahoma statute instructs the State Water Resources Board to reserve to the area of origin sufficient water to take care of its present and reasonable future needs.

Because of the "watershed limitation" of eastern riparian law, areas of origin in the East might be able to obtain even greater protection than their western counterparts. In 1905, when New York City sought to obtain water from the Catskill Mountain area, it was forced by the state legislature to compensate residents of the area of origin for every conceivable claim that might arise from these projects. The city paid not only for the value of the property, buildings, and equipment actually taken, but also for business and wage losses both to riparians whose property was taken and to nonriparians in nearby areas who were adversely affected by the projects.[47] Since under riparian law diversion rights are inherent in landownership, the proponent of a major interbasin diversion in the East would have to (1) convince the legislature in the state of origin to modify traditional watershed limitations, and (2) compensate at least the riparian owners on source rivers for their often inchoate and unrecorded riparian rights. Especially in eastern nonpermit states, these barriers to major interbasin transfers could prove insurmountable.

In the Great Lakes area, governmental units have banded together to protect themselves against undesirable transfers of water out of the Basin. The governors of eight Great Lakes states and the premiers of the Canadian provinces of Ontario and Quebec signed the Great Lakes Charter in 1985. Under this charter, the states and provinces will provide prior notice and consultation of any new or increased diversion exceeding 5 million gallons per day average in any 30-day period. The charter also provides that "diversions of Basin water resources will not be allowed if individually or cumulatively they would have significant adverse impacts on lake levels, in-basin uses and the Great Lakes Ecosystem." Four states (Illinois, Indiana, Minnesota, and Ohio) have gone beyond this legally-unenforceable charter. They have enacted statutes prohibiting diversion of Great Lakes water through their states without approval of all four states and the International Joint Commission. By treaty between the United States and Canada, the IJC—composed of three commissioners from each nation—must approve all prospective projects that involve significant diversions of water from the Great Lakes Basin. In 1986 Congress took a hand in the matter of Great Lakes diversions. As part of the Water Resources Development Act, it banned all diversions outside the basin without the unanimous approval of Great Lakes governors.

STATE ANTIEXPORTATION STATUTES

These statutes form another obstacle to major interbasin transfers. Many states have enacted statutes directly or indirectly prohibiting or regulating the export of water outside the state.[10,14] Some antiexportation statutes prohibit any exportation of water; others bar exportation for certain purposes (e.g., coal slurry pinelines); others ban exportation without legislative or administrative approval; still others allow exportation only if the recipient state allows reciprocal transfers. Until 1982, the U.S. Supreme Court had not considered the constitutionality of antiexportation statutes in light of the so-called "negative Commerce Clause": Under Article I, Section 8, the Commerce Clause of the U.S. Constitution, the federal government has the power to regulate interstate commerce and prevent a state from imposing an "undue burden" on commerce. Prior U.S. Supreme Court decisions had struck down as "economic isolationism" state prohibitions or restrictions on exportation of natural resources, e.g., minnows and minerals. Is water an "article of commerce" like these? Are antiexportation statutes inherently illegal?

In *Sporhase* v. *Nebraska*,[48] the Supreme Court considered a Nebraska statute forbidding the export of any state water, including groundwater, unless the recipient state allowed export of its water to Nebraska. First, groundwater is indeed an article of commerce subject to federal regulation and Commerce Clause analysis. The Court then went on to determine the constitutionality of the Nebraska antiexportation statute. Although antiexportation statutes are inherently illegal where other natural resources are concerned, water antiexportation statutes are suspect but not inherently illegal because of the importance of water to public health—basically a state concern—and the traditional federal deference to state water law. Because of the preferred status of state water regulation, water antiexportation statutes will be upheld if reasonable and "narrowly drawn," i.e., no broader than necessary to accomplish the purpose of the statute. For example, a Nebraska-type statute might be upheld "if it could be shown that the State as a whole suffers a water shortage, that the interstate transportation of water from areas of abundance to areas of shortage is feasible regardless of distance, and that the importation of water from adjoining States would roughly compensate for any exportation to those States." Even an absolute ban on water exportation could be upheld where "a demonstrably arid State [is] able to marshall evidence to establish a close means-end relationship between . . . a total ban on the exportation of water and a purpose to conserve water."

The Nebraska statute, however, was struck down because it was too loosely drawn. Despite Nebraska's strict regulation of intrastate transfers of groundwater from critical aquifers, the Nebraska statute burdened interstate commerce because "even though the supply of water in a particular well may be abundant, or perhaps even excessive, and even though the most beneficial use of that water may be in another State, such water may not be shipped into a neighboring State that does not permit its water to be used in Nebraska." Consequently, in order to pass constitutional muster, Nebraska could either redraft its statute or restrict the exportation of groundwater through its normal allocation process. Where water is concerned, a state "may favor its own citizens in times of shortage."

The *Sporhase* Rule was applied by a federal district court in *El Paso* v. *Reynolds*.[49] El Paso wanted to import water from wells it had dug across the border in New Mexico, but found itself frustrated by a New Mexico statute banning exportation of groundwater. The court struck down the New Mexico statute as unconstitutional because New Mexico allowed long-distance intrastate transfers of groundwater and because New Mexico had not shown a pressing need for the water. New Mexico then repealed its groundwater embargo and replaced it with an administrative permit system. Several other states have weakened their antiexportation statutes in response to the Sporhase decision.

Professor Trelease believes that *Sporhase* will have limited applicability because it does not govern allocations of water under interstate compacts, equitable apportionment decrees, and congressional apportionments.[50] Other commentators predict that Sporhase, by making groundwater an "article of commerce," will usher in a new era of federal groundwater regulation.[22] Professor Utton argues that the Supreme Court took the wrong approach in *Sporhase*: instead of adhering to a free market solution to distributing water from interstate aquifers, it should have promoted conjunctive management by applying its interstate surface water allocation principle—equitable apportionment—to interstate groundwater.[51] These are three of the varied reactions elicited by *Sporhase*. One certainty is that its judicial, legislative, and administrative progeny will bear careful watching.

Chapter 8
DRAINAGE LAW

Drainage law consists of those principles that govern a landowner's rights to repel water at the boundaries of his land and expel it once it has entered. Consistent with the legal distinction between diffused surface water and water in watercourses, drainage law applies different rules to drainage of diffused surface water and drainage of overflow from watercourses. As previously pointed out, state courts frequently disagree about whether floodwater, springs, and swamps are diffused surface water or watercourses. Thus, a court's classification of waters will strongly affect its decision about drainage rights as well as rights to divert water for beneficial purposes. Drainage law also distinguishes between drainage of surface water and groundwater.

DRAINAGE OF DIFFUSED SURFACE WATER

Three different rules exist as far as drainage of diffused surface water is concerned: (1) the common enemy rule, (2) the natural flow rule, and (3) the reasonable use rule.[52]

In common enemy rule states, a landowner is allowed to do whatever is necessary to repel or expel diffused surface water regardless of damage to his neighbor's land. However, the potential harshness of the common enemy rule is tempered by exceptions that qualify the rule in most common enemy states: (1) a landowner may not dispose of water by an artificial channel (e.g., drain, culvert, ditch) if his neighbor is injured; and (2) a landowner may not drain into a natural watercourse so as to obstruct its flow or overtax its capacity to another's injury. Courts in common enemy states have also infused elements of "reasonableness" into drainage cases.

Nevertheless, a number of states that once followed the common enemy rule have abandoned it in favor of the reasonable use rule. Indiana, Maine, and Montana still apply an unmodified common enemy rule, while Arkansas, Connecticut, Nebraska, South Carolina, and the District of Columbia subscribe to a modified rule.[10]

The natural flow rule, in its pure form, is as radically antidevelopment as the common enemy rule is prodevelopment. Under the natural flow rule, a landowner may not interfere with natural drainage patterns where this would affect neighboring landowners. Like common enemy states, some natural flow states have made exceptions for agricultural drainage and drainage into natural watercourses where capacities are not exceeded. Other natural flow states have made exceptions for drainage in areas where dwellings are close together. Reasonableness is also a criterion in many natural flow decisions. But the following states still appear to follow an unmodified natural flow rule: Delaware, Florida, Georgia, Idaho, Kansas, Louisiana, Michigan, Mississippi, Nevada, New Mexico, Tennessee, Texas, Vermont, and West Virginia.[10]

The reasonable use rule has become the majority rule. Under the reasonable use rule, a landowner is privileged to make reasonable use of his land even though the flow of diffused surface water is altered and a neighbor is damaged. Only when harm to a neighbor becomes unreasonable is the landowner liable for damage occasioned by his drainage project. Professor Beck has listed some of the questions bearing on reasonableness: Is there a reasonable need for the land development? Has due care been taken to prevent injury? Has the natural drainage pattern been followed as much as possible? Is the artificial drainage system reasonably feasible? Do the benefits of the project outweigh its harm?[52] Once again, the chief advantage of the reasonable use rule is its flexibility; its major disadvantage is its unpredictability.

Nowadays, major drainage projects are constructed by public drainage institutions such as drainage districts or municipalities. All drainage districts and municipalities depend on state statutes for their existence and powers. As a result, there is a great deal of variation in public drainage law from state to state. In many states there are several different types of entities involved with drainage, e.g., drainage districts, reclamation districts, flood control districts, irrigation districts, soil and water conservation districts, water management districts, levee districts, mosquito control districts, and sewerage authorities. One of the factors accounting for this diversity in public districts is a trend from single-purpose to multipurpose districts without abolishing the old districts.[52] Moreover, the powers of state agencies as to drainage vary among states. Some state engineers have authority over drainage; others do not. Some state environmental protection agencies have power to control urban stormwater runoff; others do not. All in all, it is difficult to generalize about the formation, powers, financing, and dissolution of public bodies dealing with drainage.

However, there are two aspects of public drainage law that deserve special mention. Essentially all states give public entities doing drainage the power of eminent domain to condemn land and water rights for drainage purposes. Second, public drainage institutions are given "sovereign immunity." This means that drainage districts or municipalities or their officials cannot be sued for their "torts" (wrongs), when acting in their official capacities, without the consent of the state legislature or state courts. In most states, suits against public officials are sharply restricted either by state tort claims laws or common law rules of governmental immunity. For example, in New Jersey a suit cannot be brought against a public drainage body unless its action has been "palpably unreasonable"—a much higher standard than ordinary reasonableness. Thus, in lawsuits against public drainage institutions the three surface water drainage rules are replaced by the generally stricter provisions of tort claims laws and court decrees.[53] This emphasizes the need for effective public regulation of drainage districts.

GROUNDWATER DRAINAGE

These cases are of two kinds: (1) where a landowner seeks to block the flow of groundwater from coming onto his premises, and (2) where a landowner's reduction of the watertable for dry development causes dewatering or subsidence of neighboring land.[52] In the first category of cases, courts generally apply the same rule that applies to drainage of diffused surface water. The second class of cases has proved more difficult for courts. Most of the dewatering and subsidence cases involve mining, although there are also a number of sewer construction cases.[54] In states that follow the "absolute ownership" of groundwater rule, there is generally no recovery of damages for dewatering or subsidence. Courts in reasonable use states will come to the same conclusion if the drainage of groundwater is not wasteful (negligent) and the drainage water is not transported away from the overlying land. In states following the "correlative rights" rule, where owners of adjacent land have a common and coequal right to percolating water, the injured neighbor usually wins, even if the mine drainage is reasonable and necessary and the mine is much more valuable than the plaintiff's property. In those western states where groundwater is subject to appropriation, the landowner cannot intercept a source of supply to streams, springs, or groundwater used by prior appropriators. The miner's only alternative is to prove that the mining activity has developed new water not tributary to any fully appropriated water source.[54] New Mexico has attempted to resolve the conflict between the prior appropriation doctrine and the needs of the mining industry by enacting a Mine Dewatering Act, which allows miners to replace water previously appropriated, e.g., to furnish a substitute water supply, to drill a deeper well for the affected user, or to compensate him for increased lift costs.

In subsidence cases there is a further complication. Adjoining land-owners owe each other absolute obligations of support, both surface support and subsurface support. A landowner must not use his land so as to cause his neighbor's land to subside, either at or below the surface. If dewatering leading to subsidence of neighboring land is a breach of the obligation to support neighboring land, then plaintiffs will recover in all cases. Many courts faced with the problem have limited the "support" theory to situations involving removal of solid materials or water mixed with solid material, but other decisions refer only to dewatering.[17] In fact, the law relating to subsidence caused by groundwater withdrawal is replete with confusion and inconsistency.[55] Arizona, California, and Texas have enacted statutes to deal with this problem.

DRAINAGE OF WATERCOURSE OVERFLOW (CHANNEL MODIFICATIONS)

The general rule is that no one has the right to obstruct or divert the flow of a natural watercourse, or to divert the water from its natural channel into another channel, if the stream modifications cause new and injurious overflows.[27] This rule applies to on-channel and off-channel impoundments, dikes, embankments, levees, and other artificial channel modifications. Thus the operator of a dam may permit floodwaters to pass over the dam in an amount equal to the inflow, but will be liable if any excess amount is discharged.[56]

However, like most drainage law rules, the "no increased overflow" rule is often modified by concepts of reasonableness. For example, temporarily accelerated flows for the purpose of generating hydroelectric power may not give rise to liability if the releases, even though harmful, are as innocuous as possible.[57] A dam operator may be liable for normal downstream flood damage if he should have anticipated the flood conditions and released water in advance so as to have room in the reservoir for storage of flood flows. An embankment builder may not be liable for damage caused by extraordinary—as opposed to merely ordinary—floods. Or, a dam owner might not be liable for backflooding of land upstream of the dam if the damage is so unusual that it could not have reasonably been foreseen.

When a dam breaks or collapses, liability depends on where the dam is located and who operates it. The federal government is statutorily insulated from liability "for any damages from or by flood waters at any place."[56] A majority of states impose liability on a dam operator in a dam-break case only where he has been negligent, i.e., has failed to use reasonable care in the design, operation, or maintenance of the dam. Other states impose "strict liability"—liability without fault—in such cases.

Because common law rules for determining a dam owner's liability are vague, and in the light of tragic dam failures such as the one at Buffalo Creek,

West Virginia, Congress has authorized the Corps of Engineers to inspect nonfederal dams and make program grants to states for inspection and repair of unsafe dams. States with approved dam safety programs may receive 50% matching grants out of $13 million per year from 1988 through 1992. The elements of an approvable state program are (1) preconstruction review; (2) construction supervision and preoperation review; (3) periodic inspection; (4) more frequent safety inspection, where necessary; (5) requirements for modifications to assure safety; (6) emergency response plans; and (7) authority to order or make necessary repairs. A National Dam Safety Board, consisting of federal and state officials, has been established to review and monitor state dam safety programs. The Corps is authorized to train state dam safety inspectors. No federal funds may be used to repair a nonfederal dam. The existing dam safety program of the Federal Emergency Management Agency should complement the Corps program.

Many states attempt to forestall flood damage problems by requiring permits for dam construction, stream encroachments, and floodplain development. Effective regulation of channel and bank modifications is especially important because these projects are often undertaken by governmental entities (municipalities or special-purpose districts) that are exempt from liability most of the time. Additionally, a casually granted permit can effectively insulate a defendant from liability. Although the general rule is that possession of a permit is no defense to a common law action, judges treat a permit as bearing on the reasonableness of a defendant's conduct. Moreover, governmental entities are exempt from liability for negligently issued permits.

Because constitutions are "fundamental law," governmental entities cannot escape liability when their channel modifications cause "takings" of property for which just compensation must be paid under the U.S. Constitution and most state constitutions.[58] Some states grant a considerably broader right to their citizens to sue for compensation when their property is merely damaged and not completely "taken."[53] When channel modifications lead to permanent overflows of land above ordinary high-water mark, compensation for a taking is necessary.[58] The "navigation servitude," which enables the United States to take without compensation riparian rights within the banks of navigable watercourses, is discussed in Part II. When the overflows are periodic, a governmental entity is liable for the value of a "flowage easement" (right to inundate part but not all of a piece of property) without having to condemn the entire parcel. Temporary flooding that is relatively minor in effect is not compensable. There is generally no liability when diversion of water from a channel washes away property by erosion, but when erosion results directly from flooding, rather than from a change in the direction of streamflow, a compensable taking occurs.

Chapter 9
WATER DISTRIBUTION ORGANIZATIONS

There are three major classes of organizations that distribute water to ultimate consumers: municipal water supply organizations, irrigation organizations, and water management districts in California.

MUNICIPAL WATER SUPPLY ORGANIZATIONS

Municipal water supply management in the United States is highly fragmented with respect to types and numbers of systems.[59] For example, water for cities and towns in California is delivered by municipal water departments, municipal utility districts, and nine other types of public agencies. Moreover, 50 special acts have created public districts that to some extent provide urban areas with water.[60] As an example of water supply fragmentation within a metropolitan complex, the Chicago metropolitan area has 349 separate water supply systems.[61] The inefficiencies caused by these atomized systems have prompted calls for consolidation and regionalization,[59] not only of water supply organizations but also of all organizations performing the three basic water services in metropolitan areas—water supply, wastewater collection and treatment, and stormwater management.[61]

Fragmentation of water systems along political lines is prevalent in the United States:

> The development and endurance of fragmented water systems reflect a strong tradition of local autonomy. Municipalities generally try to avoid becoming

dependent for direct services on neighboring supply systems. Instead, each municipality almost always establishes its own system or franchises an investor-owned system.[59]

In addition, three different types of water distribution systems coexist in most states: municipally-owned systems, investor-owned systems, and special districts and authorities.

Municipally-owned systems have several advantages over investor-owned systems. First, their rates and services are either exempt from regulation or subject to a lesser degree of regulation by state public utilities commissions. Second, "municipally-owned systems pay neither income taxes on revenues nor property taxes on property within municipal borders."[59] Third, municipally-owned systems can finance capital improvements more readily because of their access to the municipal bond market. On the other hand, because the revenues from municipal water systems are typically diverted to other municipal purposes, and because local politicians are loathe to raise water rates, municipally-owned systems are often poorly maintained.

When evaluating investor-owned systems, one must distinguish between large and small systems.[59] Most investor-owned systems are small, underfinanced, and inefficient. In states like New Jersey and New York there has been a trend toward municipal systems taking over small, failing investor-owned systems. Conversely, large investor-owned systems often perform better than municipally-owned systems.

According to Professor Gellis:[59]

> Special districts and public authorities theoretically enjoy the best of two worlds. As quasi-municipal entities, they receive favorable tax treatment. Almost all states exempt these systems' property, wherever located, from taxation. As semi-autonomous public bodies, they can manage their affairs more efficiently and with less red tape than most government agencies. . . . Quasi-municipal entities also offer the advantage of structural flexibility, since jurisdictional lines can be drawn to reflect regional, rather than local needs.

Nevertheless, many Americans are suspicious of these entities because of their lack of accountability and political legitimacy.

IRRIGATION ORGANIZATIONS

Both private and public organizations furnish water for irrigation in the West.[60] The private organizations range from associations of relatively few farmers to large nonprofit corporations and commercial irrigation companies. The public organizations are usually irrigation districts. Irrigation organizations "acquire, hold, and exercise appropriative rights in order to provide water for land which they were organized to serve."[27]

The most popular type of private irrigation organization is the mutual water company, "a non-profit corporation that owns diversion or storage works and delivers water at cost to users who own its stock, and that derives its operating funds from assessments levied against the stockholders."[62] Stock is divided among the stockholders based on proportionate shares of irrigable acreage. Ownership of stock in a mutual company may impair the ordinarily unlimited transferability of appropriative water rights. Articles of incorporation frequently prohibit transfers of stock apart from the irrigated land or without the consent of the corporation. State statutes and court decisions may also limit transfers of stock apart from the land. One of the reasons for restrictions on the permanent and temporary (e.g., leasing, pledging) transfers of mutual company stock is to preclude regulation of a mutual company as a public utility.

All 17 western states have irrigation district laws, although some districts are called water conservation, water improvement, or reclamation districts.[62] These state statutes differ greatly as to methods of forming irrigation districts. Once formed, irrigation districts have the authority to issue bonds to fund construction of irrigation works. Bonds are financed by assessments on land within the district; assessments are based either on value of irrigation benefits received or some ad valorem formula based on land area or property tax. Courts will look carefully at ad valorem assessments to determine whether they bear some relationship to benefits. Irrigation districts also raise revenues by tolls or charges for water delivered.[60] Many irrigation districts in the modern West "are involved in other lucrative activities such as hydroelectric power generation, operation of recreation facilities, drainage, flood control, sanitation, and municipal and industrial water supply."[10]

CALIFORNIA'S WATER MANAGEMENT DISTRICTS

In California, state level involvement in groundwater management is limited to information collection and dissemination.[63] The situation in most parts of the state is that "users pump without restraint until overdraft occurs and then resort to the courts for allocation of an insufficient resource. In fact, until overdraft begins there is little else that can be done because there are generally no entities with pre-litigation management authority."[64] Once a basin has been adjudicated, management is placed in the hands of a court-appointed watermaster. Most California basins, however, have not been adjudicated, and there "groundwater management is nothing more than the cumulative decisions of individual pumpers."[63]

Although over 900 special districts (of 138 different types) perform water distribution functions in California, effective district management of groundwater is evident in only a few basins. The Orange County Water District

(OCWD) is generally presented as a model of district management. Lacking regulatory powers, the OCWD achieves conjunctive management through a complicated set of taxes and incentives.[64] During overdraft periods, the OCWD is authorized to levy pump taxes, the proceeds of which are used to purchase imported surface water for aquifer recharge. The Water District also purchases imported water for direct use by consumers. With regard to direct sales, OCWD imposes basin equity assessments that equalize the costs of groundwater and surface water. First, OCWD sets the percentage that groundwater is to contribute to the total amount of water used in the district.

> Each producer of groundwater is given a production limit based on the percentage, and must make up the difference through purchases of imported water. Producers who pump more than their production requirement must pay a per acre-foot assessment for the excess produced. Those who pump less than their allotment are then paid out of the fund. Thus, there is no economic incentive to pump rather than purchase water from supplemental sources.[64]

The District is also authorized to levy ad valorem taxes on all property owners.

OCWD has prevented groundwater mining "not by reducing withdrawals but by increasing replenishment."[65] Can this strategy continue in light of California's decreased share of Colorado River water due to increased diversions for the Central Arizona Project?

REFERENCES FOR PART I

1. Clark, R. E. "Classes of Water and Character of Water Rights," in *Waters and Water Rights*, R. E. Clark, ed. (7 vols.; Indianapolis, IN: Allen Smith Co., 1967–1978), I, p. 300. The seven-volume treatise is hereafter referred to as "Clark."
2. Davis, C. "The Right to Use Water in the Eastern States," in Clark, VII, p. 154.
3. Davis, C. "Introduction to Water Law of the Eastern States," in Clark, VII, p. 6.
4. Clark, R. E. "Western Ground-Water Law," in Clark, V, p. 419.
5. Corker, C. E. "Groundwater Law, Management and Administration," National Water Commission, NTIS PB205 527 (1971), pp. 56–59.
6. Champion, W. M. "Ground Water Rights," in *Water Rights in the Nineteen Western States*, W. A. Hutchins (completed by H. H. Ellis and J. P. DeBraal) (3 vols.; Washington, D.C.: U.S. Government Printing Office, 1971–1977), II, p. 639. The three-volume treatise is hereafter referred to as "Hutchins."
7. *The Random House Dictionary of the English Language* (New York, 1967).
8. Sax, J. L. and R. H. Abrams, *Legal Control of Water Resources* (St. Paul, MN: West Publishing Company, 1986), p. 154.
9. Ausness, R. C. "Water Use Permits in a Riparian State: Problems and Proposals," *Kentucky Law Journal* 66:191–265 (1977), p. 197.
10. Getches, D. H. *Water Law in a Nutshell* (St. Paul, MN: West Publishing Company, 1984) p. 58.
11. Miller, D. W., M. S. Heath, Jr., and R. E. Sneed, "Riparian Rights to Streamflow and Their Application in Eastern States," in *Water Resources Law* (St. Joseph, MO: ASAE, 1986), p. 68.
12. Davis, P. N. "Eastern Water Diversion Permit Statutes: Precedents for Missouri?", *Missouri Law Review* 47:426–470 (1982), p. 434.
13. Tarlock, A. D., "Introduction to Symposium on Eastern Water Rights," *William and Mary Law Review* 24:535–545 (1983).
14. Sherk, G. W., "Eastern Water Law," in *Natural Resources & Environment* (Chicago, IL: American Bar Association, Section of Natural Resources Law, 1986), Vol. I, No. 4, 7ff.
15. Ausness, R. C. "Water Rights Legislation in the East: A Program for Reform," *William and Mary Law Review* 24:547–590 (1983), p. 555.
16. 371 So. 2d 663, *rehearing den* (1979), *cert. den.* 444 U.S. 965 (1979). The Florida Water Resources Act can be found at §373.016 Fla. Stat. et seq.
17. Meyers, C. J., and A. D. Tarlock, *Water Resources Management*, 2nd ed. (Mineola, NY: The Foundation Press, 1980), p. 197.
18. Clark, R. E. "Introduction to Water Law of the Western States," in Clark, V, pp. 7–8.
19. Clark, R. E. "The California Doctrine," in Clark, V, pp. 234–235.

20. Clark, R. E. "The Colorado Doctrine," in Clark, V, p. 66.
21. Trelease, F. J. "Federal-State Relations in Water Law," National Water Commission, NTIS PB203 600 (1971).
22. Gould, G. A. "Water Law In 1986: Selected Issues," in *Water Resources Law* (St. Joseph, MO: ASAE, 1986), p. 3.
23. Radosevich, G. E., and D. R. Daines. "Water Law and Administration in the United States of America," in *Proceedings of the International Conference on Global Water Law Systems* (Fort Collins: Colorado State University, 1976), p. 25.
24. *State Dept. of Parks* v. *Idaho Dept. of Water Administration*, 96 Idaho 440, 530 P. 2d 924 (1974).
25. Hutchins, W. A. "Background and Modern Developments in State Water Rights Law," in Clark, I, pp. 88–91.
26. Rice, L., and M. D. White, *Engineering Aspects of Water Law* (New York: John Wiley & Sons, 1987), pp. 25–26.
27. Hutchins, I, p. 277.
28. Tarlock, A. D. "An Overview of the Law of Groundwater Management," *Water Resources Research* 21:1751–1766 (1985), p. 1752.
29. Lukas, T. N. "When the Well Runs Dry," *Boston College Environmental Affairs Law Review* 10(2):445–502 (1982), p. 484.
30. Hanks, E. H., and J. L. Hanks. "The Law of Water in New Jersey: Groundwater," *Rutgers Law Review* 24:621–671 (1970), pp. 638–639.
31. 33 Cal. 2d 908, 207 P. 2d 17 (1949).
32. Getches, D. H. "Controlling Groundwater Use and Quality: A Fragmented System," *Natural Resources Lawyer* 17:623–645 (1985), p. 627.
33. Weatherford, G., et al., "California Groundwater Management: The Sacred and the Profane," *Natural Resources Journal* 22:1031–1043 (1982), p. 1035.
34. *Cappaert* v. *United States*, 426 U.S. 128, 96 S. Ct. 2062, 48 L. Ed. 2d 523 (1976).
35. 438 U.S. 696, 98 S. Ct. 3012, 57 L. Ed. 2d 1052 (1978).
36. 43 USC §666.
37. Tarlock, A. D., "The Endangered Species Act and Western Water Rights," *Land and Water Law Review* 20:1 (1985).
38. 568 F. Supp. 583 (1983).
39. Corker, C. E. "Water Rights in Interstate Streams," in Clark, II, p. 294.
40. Ellis, H. H., and J. P. DeBraal. "Interstate Dimensions of Water Rights," in Hutchins, III, p. 71.
41. Utton, A. E., "In Search of An Integrating Principle For Interstate Water Law: Regulation versus the Market Place," *Natural Resources Journal* 25: 985–1004 (1985).
42. Tarlock, A. D. "The Law of Equitable Apportionment Revisited, Updated, and Restated," *University of Colorado Law Review* 56:381–411 (1985).
43. Meyers, C. J., A. D. Tarlock, et al., *Water Resource Management*, 3rd Ed. (Mineola, NY: The Foundation Press, 1988), pp. 987–1007.
44. Muys, J. C. "Interstate Water Compacts," National Water Commission, NTIS PB 202 998 (1971), p. S-1.
45. Utton, A. E. "International Streams and Lakes," in Clark, II, pp. 404–411.
46. Waite, G. G. "International Law Affecting Water Rights," in Hutchins, III, p. 119.

47. Johnson, R. W. "Major Interbasin Transfers: Legal Aspects," National Water Commission, NTIS PB 202 619 (1971), p. 87.
48. 458 U.S. 941, 102 S. Ct. 3465, 73 L. Ed. 2d 1254 (1982).
49. 563 F. Supp. 379 (1983).
50. Trelease, F. J. "State Water and State Lines: Commerce in Water Resources," *University of Colorado Law Review* 56:347–379 (1985), p. 350.
51. Utton, A. E. "Alternatives and Uncertainties in Interstate Groundwater Law," *Water Resources Research* 21: 1767–1770 (1985).
52. Beck, R. E. "The Law of Drainage," in Clark, V, pp. 477–518.
53. Goldfarb, W., and B. King. "Urban Stormwater Runoff: The Legal Remedies," *Real Estate Law Journal* 11:3–46 (1982), p. 40.
54. Clyde, E. W. "Mining and Land-Development Interference with Ground Water," in Clark, V, pp. 460–474.
55. Quinlan, J. F. "Legal Aspects of Sinkhole Development and Flooding in Karst Terranes," *Environmental Geology and Water Science* 8:41–61 (1986).
56. Binder, D. "Risk and the Calculus Of Legal Liability In Dam Failures," in *Safety of Dams*, National Research Council (Washington, D.C.: National Academy Press, 1985), p. 108.
57. Davis, C. "Impoundment and Artifical Development of Waters in the East," in Clark, VII, pp. 198–199.
58. Harnsberger, R. E. "Eminent Domain and Water," in Clark, IV, p. 153.
59. Gellis, A. J. "Water Supply in the Northeast: A Study in Regulatory Failure," *Ecology Law Quarterly* 12:429–479 (1985), p. 436.
60. Swensen, R. W. "Local and Private Water Distribution Agencies," in Clark, IV, p. 447.
61. "Water Policies for the Future," National Water Commission, U.S. Government Printing Office (1973), p. 443.
62. Trelease, F. J. *Cases and Materials on Water Law*, 2d ed. (St. Paul, MN: West Publishing Co., 1974), p. 612.
63. Smith, Z. A. "Centralized Decisionmaking in the Administration of Groundwater Rights: The Experience of Arizona, California and New Mexico and Suggestions for the Future," *Natural Resources Journal* 24:641–688 (1984), p. 660.
64. De Lambert, D. A. "District Management for California's Groundwater," *Ecology Law Quarterly* 11:373–400 (1984), p. 387.
65. Smith, P. K., "Coercion and Groundwater Management: Three Case Studies And A 'Market' Approach," *Environmental Law* 16:797–882 (1986), p. 834.

PART II

Water Resources Development and Protection

The law of water diversion and distribution is primarily state law. However, because the federal government is the major developer of water resources in the United States, and because of the traditional federal guardianship over navigable waters, the law regarding development of water resources and protection of them against unwise development is predominantly federal law. States utilize their taxing and bonding powers to finance water resources development, and important state environmental protection statutes apply to these projects. But state laws vary to such an extent that this part focuses almost exclusively on federal law.

The federal government exercises powers delegated to it by the states and enumerated in the U.S. Constitution. "The Constitution does not contain the word water, yet the founding fathers provided for a strong nation, and the powers they gave to the central government have enabled it to engage in many water related activities and to undertake the most extensive program of water resources development in the world."[1] The most important source of federal power over water resources is Article I, Section 8 of the U.S. Constitution, the so-called "commerce clause." giving Congress the power to "regulate Commerce . . . among the several States." By 1865, the U.S. Supreme Court, interpreter of the Constitution, had held that "commerce" includes "transportation," which in turn includes "navigation," and that the power to regulate navigation includes the control of navigable waters for the purposes of navigation. The Court's broad interpretation of the "navigation power" has continued:

> The power to control navigation and navigable waters includes the power to construct obstructions that destroy the navigable capacity of the waters and prevent navigation. It also includes the power to protect navigable capacity by preventing diversions of the river and even its nonnavigable tributaries, or by preventing obstructions by bridges or dams and the power to construct flood control structures on the navigable waters, their nonnavigable tributaries or even on the watersheds. Powers to prevent obstruction in turn leads to powers to license obstructions. The power to obstruct leads to the power to generate electric energy from the dammed water.[1]

The federal navigation power has become a broad authority for federal water resources project development and licensing. In recent years it has also become the foundation for a wide range of federal environmental protection statutes that can be utilized to protect water resources against insensitive federal development or licensing of private development.

Liberal judicial definitions of "navigability" have further widened the federal navigation power. For these purposes, a navigable waterbody is one which has been in the past, is now, or could with reasonable improvements be used as part of a continuous interstate waterway for commercial purposes; for example, floating sawlogs (see Chapter 11). Nonnavigable portions of navigable waterbodies and nonnavigable tributaries that affect the navigable capacity of navigable waters are within the scope of the navigation power. "Theoretically, therefore, few waters in the United States are immune from the navigation power."[2] Moreover, the U.S. Supreme Court has upheld the Federal Energy Regulatory Commission's authority to require a federal license for a pumped storage project on a nonnavigable stream which did not affect downstream navigation but which generated power to be transmitted in interstate commerce.

Far-reaching as congressional authority over water resources development may be under the navigation power, it must be kept in mind that Congress is not required to exercise this authority. Congress does not have to authorize particular dams or require licenses for particular projects. Specific water resources development statutes are a function of the political process. The virtually unlimited navigation power simply means if Congress chooses to legislate in the water resources development area it probably cannot be stopped by arguments that it lacks the constitutional authority to do so.

The federal navigation power is often confused with the federal "navigation servitude." The latter is "a shorthand expression for the rule that in the exercise of the navigation power certain private property rights may be taken without compensation."[2]

> Any private property right whose exercise or value depends on the use or presence of navigable waters is burdened with the "navigation servitude." This means that if it is taken, destroyed or impaired by the federal govern-

ment's exercise of powers over navigable waters, the government is not constitutionally required to pay its value or the damages, although it would have paid full compensation if it had taken or destroyed the right by the exercise of any other power. The theory is that government takes nothing; the title to the property always contained this defect or limitation by which it might be terminated.[1]

The navigation servitude is synonymous with this "no compensation rule." We have seen in Part I, Chapter 2, that riparian owners in all states have rights to wharf out, have access to the watercourse, fish, alter or protect shoreline areas, and recreate on all or part of the surface. These rights may be defeated, without compensation, by the navigation servitude. For example, a riparian owner would be required to destroy a wharf that impedes navigation; an oyster fisherman would have no remedy if his privately-owned oyster beds are destroyed by Corps dredging; or an upstream dam owner would lose the water power value of his dam if backflow from a downstream federal dam diminished his power head. Moreover, water diversion rights perfected under state law may also be defeated by the federal navigation servitude. Appropriators whose headgates are closed because of the need to maintain instream flows by virtue of federal regulatory water rights (Part I, Chapter 6) may not be entitled to compensation, depending upon judicial interpretations of the scope of the federal navigation servitude (see Chapter 16).

However, the federal navigation servitude is limited in two ways. First, although the navigation power applies to nonnavigable tributaries of navigable waterways, the navigation servitude generally does not. Thus, where backflow from a federal dam destroys the power potential of a dam located on a nonnavigable tributary, the dam's owner is entitled to compensation. Second, the navigation servitude extends only to the ordinary high-water mark of navigable waterbodies. When lands above the high-water mark ("fast lands") are taken in the exercise of the navigation power, just compensation must be paid.[3] By federal statute, the fair market value of fast lands is based on "all uses to which such real property may reasonably be put, including its highest and best use, any of which uses may be dependent upon access to or utilization of such navigable waters."[4] This statute has not been tested in the courts, but if taken at face value, it means that fast land values may include factors that would be noncompensable if the land were located below the high-water mark.

The foregoing discussion of the navigation power and navigation servitude has been a necessary introduction to Part II, which deals with the roles of the federal government in water resources development and protection. In Part I, two of the three criteria relevant to water resource development projects—efficiency and equity—were emphasized. The third "e"—environmental protection—will now be accorded its rightful position in this triad of values by which water resources development must be measured.

Chapter 10

THE FEDERAL GOVERNMENT AS PROJECT DEVELOPER

BUREAU OF RECLAMATION

The Bureau, located within the Interior Department, operates in the 17 western states and Hawaii.

> Historically, the primary function of the Bureau was to facilitate the settlement and development of Western lands through the provision of irrigation water. Today most projects are multipurpose and some do not have an irrigation component at all. However, irrigation remains the primary purpose and, in 1969 Bureau projects served approximately 8.6 million acres, slightly more than a fifth of all the irrigated land in the United States in that year.[5]

The Bureau also has planning responsibilities regarding water from the Colorado River, as well as control of various loan and grant programs.

In the fall of 1987, the Bureau announced that instead of constructing large projects, as it had in the past, it will concentrate on managing the existing projects and assuring water conservation and environmental protection. The Bureau will be completely reorganized, and its 8000-member staff will be cut by as much as 50% over the next decade. In addition, its home office will be moved from Washington, D.C. to Denver.

Reclamation law is a complex system of statutes, administrative actions, and court cases, based on the Reclamation Act of 1902. In 1982, Congress made profound changes in reclamation law when it passed the Reclamation Reform Act of 1982. These changes can be better understood if certain aspects of earlier reclamation law are explained.

76

"The basic principle of the 1902 Reclamation Act was that the United States would build irrigation works from the proceeds of public-land sales in the . . . arid western states."[6] Irrigators were required to repay operation and maintenance costs each year, and the costs of construction in full within 10 years, but received a governmental subsidy in that no interest accrued on the construction costs. These repayments were intended to establish a revolving fund to finance future reclamation projects. But the reclamation fund has never been self-supporting:

> Substantial construction monies are appropriated from the general treasury; the ten year interest-free time for repayment by irrigation users has in practice been extended to fifty years; and irrigation users—who pay only what they are deemed able to afford—are subsidized in large part by power and municipal users.[6]

Thus, the 1902 Act's irrigation subsidy has been enhanced by the Bureau's forbearance and increasing sales of project water to power, municipal, commercial, and industrial users. Moreover, recreation, fish and wildlife, flood control, and navigation costs of a reclamation project are either partially or fully exempt from repayment because these aspects are viewed as national benefits. This further decreases the amount of project costs that irrigators are called upon to repay. The repayment obligation is included in a contract between the Bureau and an irrigation district. Irrigation districts operate the distribution systems, furnish water to irrigators, and tax lands within the district to repay construction costs and maintenance expenses.

The congressional supporters of the 1902 Reclamation Act intended its subsidies to promote family farming in the West. In order to discourage monopolization, the act provided that water could not be sold for use on more than 160 acres—expanded by Bureau practice to 320 acres for a husband and wife—in any individual private ownership, and that the user had to be a resident or occupant of the land. An exception was made to the acreage limitations to facilitate the breakup of large farms. Project water at subsidized rates (i.e., no interest demanded and ability to pay considered) could be used to irrigate land in excess of the limits if the owner agreed, in a recordable contract with the Bureau, to sell the "excess lands" within 10 years for their appraised value excluding prospective irrigation by project water. Excess lands sold for a speculative price could not be irrigated with project water.

These intended antimonopoly safeguards of the Act were generally ineffective. The Bureau virtually ignored the perhaps unenforceable residency requirement. Acreage limitations were circumvented by legal subterfuges, statutory exemptions, and liberal administrative interpretations. Critics of federal reclamation policy charged that the Bureau had perverted the "agrarian democracy" ideal of the Reclamation Act and turned the reclamation program into a giveaway to large California farming organizations.[7]

Reclamation farmers, on the other hand, argued that residency requirements and acreage limitations were anachronisms. Congress refused to modify the Reclamation Act until a federal circuit court ordered the Bureau to enforce the Act as written.

The 1982 amendments abolished the residency requirement. In addition, the acreage limitations have been loosened. Project irrigation water cannot be delivered to a "qualified recipient" (an individual or "any legal entity established under State or Federal law which benefits twenty-five natural persons or less") for use in the irrigation of land in excess of 960 acres "of Class I lands or the equivalent thereof . . . whether situated in one or more districts." Larger organizations, called "limited recipients," have an acreage limitation of 640 acres of Class I land or their equivalent. However, subsidized rates for the irrigation of excess lands have been eliminated, and for new recordable contracts the period during which excess lands may receive project water has been reduced to a maximum of five years. For these five years, full cost, including interest, must be paid by qualified recipients for irrigating with project water landholdings in excess of 960 acres of Class I land or their equivalent, and by limited recipients for over 320 acres or their equivalent. The ownership and full-cost pricing limitations do not apply where: (1) districts have fully repaid construction costs, including prepayment where allowed by existing contract with the Bureau; (2) the irrigator is temporarily provided with unmanageable oversupplies of water; (3) the land is acquired involuntarily, e.g., by foreclosure or inheritance; and (4) the excess lands are "isolated tracts found . . . to be economically farmable only if they are included in a larger farming operation but which may . . . cause it to exceed such ownership limitations."

One of the most common subterfuges for evading the Reclamation Act's acreage limitations has involved leasing. Farmers and agricultural corporations have created "paper farms" by leasing 320-acre farms from others in order to receive subsidized project water. The 1982 amendments validate leasing, but require the lessee to pay full cost for water applied to leased land where the lessee already farms 960 acres or more.

Up to a point, a district can choose whether or not to be covered by the new acreage and full-cost limitations. The 1982 amendments apply only to new and amended contracts between districts and the Bureau. Existing contracts are covered by prior law. Should a district amend its contract to conform to the new law? The advantages of amendment would be more lenient acreage limitations and application of the new "equivalency principle"; the major disadvantage would be the loss of 10 years delivery of subsidized water for excess lands under recordable contract. All other things being equal, an irrigation district, including very large farms, might choose to remain under the old law. But there is an additional penalty for refusing to amend a contract. Beginning in 1987, large farming operations leasing over 320 acres of land in nonamending districts were to lose the subsidy rates attached to the additional leased lands. This so-called "hammer clause" was intended to place farmers in amending and nonamending districts on the same footing with regard to subsidized water for leased lands above the acreage limitations.

The hammer clause has proved a highly contentious issue. After years of political skirmishing, the Bureau promulgated regulations that would permit "farm management arrangements" in which farmers could operate an unlimited number of small farms receiving subsidized water as long as the farms were legally owned by others. Farm management arrangements are distinguished from leases in the regulations. Critics of the new rules contend that these regulations condone the creation of paper farms and invite circumvention of the Reclamation Act.

Even though most reclamation law disputes are governed by the federal Reclamation Act or the bureau-district or district-user contracts, difficult questions of federal-state relations remain unanswered by the 1982 amendments. For example, the U.S. Supreme Court has held that the Bureau, in appropriating water for reclamation projects, must comply with state law just as any other appropriator, and that the state can impose reasonable conditions on the Bureau (e.g., protection of local water users or instream flow maintenance) as long as there is no conflict with federal law.[8] In its opinion, the Court suggested that federal condemnation of existing water rights is also controlled by state law. What if state law prohibits the Bureau from condemning water rights for a project? Moreover, the High Court declared that, where the Reclamation Act is silent, "once the waters [are] released from the dam, their distribution to individual landowners would again be controlled by state law." To what extent is a right to receive project water saleable if water rights are saleable under state law? Bureau policy on voluntary transfers has been informal and inconsistent.[9] Questions such as these, along with ambiguities in the Reclamation Act amendments, will dictate the future course of reclamation law.

ARMY CORPS OF ENGINEERS

The Corps of Engineers of the Department of the Army is the oldest of the federal agencies with water resource programs. Corps functions include the investigation, design, construction, operation, and maintenance of works for navigation, flood control, beach erosion control, hydroelectric power generation, municipal and industrial water supply, irrigation, water quality control, recreation, fish and wildlife conservation, and hurricane protection. The Corps also performs water supply and pollution control planning. These functions were conferred on the Corps at different times and by different statutes ("enabling acts"), most of which set out what the Corps must do in order to procure congressional authorizations and appropriations for particular public works projects. Although the details of this authorization-appropriation process differ with the specific statutes involved, they are sufficiently similar that a composite procedure can be presented:[10,11]

Step 1. *Congressional Authorization of a Reconnaissance Report.* The authorization may appear in an omnibus rivers and harbors and flood control bill, in a separate water development bill, or in a resolution adopted by either the House or Senate Committee on Public Works. Congressional authorization is not necessary for small or emergency projects.

Step 2. *Congressional Appropriation for a Reconnaissance Report.* Appropriations for authorized reports are usually made in a lump sum, from which the Chief of Engineers allocates the funds to specific reports. If the Chief does not make funds available for a project investigation within eight years of authorization, the reconnaissance report is automatically deauthorized.

Step 3. *Reconnaissance Report Preparation.* The District Engineer (the Corps is organized into 11 divisions and 36 districts) performs, at federal expense, a reconnaissance study of the water resource problem in order to identify potential solutions in sufficient detail to enable the Corps to determine whether planning should continue. The reconnaissance study must include a preliminary analysis of the federal interest, costs, benefits, and environmental impacts, and an estimate of the costs of preparing a feasibility report. No reconnaissance study may last more than 18 months. Public participation is encouraged to assist in problem identification and selection of remedial options.

Step 4. *Feasibility Report Preparation.* The Chief transmits a "Statement of Findings," based on the reconnaissance report, to the House and Senate Committee on Public Works. The statement must identify "substantial public controversy" over a project, if it exists. If it does not, the Corps can develop its feasibility report without further legislative action. If there is substantial controversy, congressional authorization of further investigations is required. The following are to be taken into account in planning Corps water resources projects: (1) national economic development (including benefits to particular regions of the nation not involving the transfer of economic activity to such regions from other regions); (2) environmental quality; (3) the well-being of the people of the United States; (4) the prevention of loss of life; and (5) the preservation of cultural and historical values. The feasibility report must describe, with reasonable certainty, the economic, environmental, and social benefits and detriments of the recommended plan and alternative plans considered by the Corps and the engineering features (including hydrologic and geologic information), the public acceptability, and the purposes, scope, and scale of the recommended plan. The feasibility report must also include the views of other federal and nonfederal agencies with regard to the recommended plan, a description of a nonstructural alternative to the recommended plan when the plan does not have significant nonstructural features, and a description of the federal and nonfederal participation in the plan. It must also demonstrate that states, other nonfederal interests, and federal agencies have been consulted in the development of the recommended plan. Preparation of the feasibility report includes development and circulation of a ''Draft Environmental Impact Statement'' and formulation of a ''Final Environmental Impact Statement'' (See Chapter 13). The Corps also prepares its ''Benefit-Cost Analysis'' at this stage. By statute, total project benefits must exceed total project costs. Benefit-cost ratios have been one of the most heavily criticized aspects of Corps project planning. Because a benefit-cost analysis is intended only for Congress, a court will not review its sufficiency independent of an Environmental Impact Statement. Nonfederal beneficiaries of a project must pay 50% of the cost of feasibility studies and all further planning and engineering for a project.

Step 5.   *Corps Review.* The Division Engineer reviews the feasibility report and, if he approves, sends it on to the Board of Engineers for Rivers and Harbors, composed of seven engineer officers of the Corps. The Board may make a site visit and hold public hearings. After its examination of previous planning, the Board makes recommendations on the feasibility report and environmental impact statement to the Chief of Engineers. The Chief conducts a further review and also sends these documents to the governors of the affected states, and in the reclamation states to the Secretary of the Interior, for a 90-day review period. He then prepares a final feasibility report, completes and files the final environmental impact statement, and issues a ''Record of Decision.''

Step 6.   *Final Project Report Submission.* The Chief sends his final feasibility report, along with any comments from affected governors or the Department of the Interior, to the Secretary of the Army. The Secretary then submits a draft of the ''letter of transmittal'' to Congress to the Office of Management and Budget (OMB) for a determination of the relationship of the

Step 7.
project to the overall program of the President. If OMB comments favorably, the final project report—including all relevant comments, the environmental impact statement, and the benefit-cost analysis—is forwarded to Congress.

Step 7. *Congressional Authorization of Project.* Congress may authorize (1) an individual project at an estimated cost; (2) a comprehensive plan of improvement and specific component parts, leaving other portions of the plan to be authorized later; or (3) a comprehensive plan and sums for plan implementation, leaving it to the Corps to choose which projects should be constructed first. No project may be authorized by Congress if more than five years has elapsed from the time the reconnaissance report was submitted. If the proposal is controversial, the Public Works Committees may hold extensive hearings. After the hearings, Congress may authorize a number of projects in a single omnibus bill.

Step 8. *Congressional Appropriation of Funds for Project.* "Authorization does not guarantee the funding and construction of projects. Congress presently has a very substantial backlog of authorized but unconstructed projects from which the Administration may select in considering which should be funded in any year . . . Some of these projects will die . . . Thirty years, or more, may pass before the first appropriation for the project."[5] On the other hand, Congress can vote to "deauthorize" a project.

Step 9. *Final Planning and Construction.* Advanced engineering and design work may require up to five years. During this time, the Corps is finalizing its cost-sharing agreements with local sponsors and acquiring project lands, including mitigation lands. Design costs are shared in the percentage applicable to the project itself. Unlike the Bureau of Reclamation, the Corps need not apply for state water rights or other state permits for its projects. Upon the completion of detailed plans and specifications, the construction phase is let out for bids and contracts are awarded. If the Corps, during actual construction, deviates substantially from the project report on which authorizations and appropriations were based, a court might enjoin construction pending issuance of a "Supplemental Environmental Impact Statement."

THE WATER RESOURCES DEVELOPMENT ACT OF 1986

Corps requests for authorizations were stalemated in Congress for a decade before a coalition of fiscal conservatives and environmentalists agreed to support the Water Resources Development Act of 1986. This statute authorized approximately 250 water projects, including projects for harbor construction, flood control, inland waterway improvement, shoreline protection, and water resource conservation and development. In addition, more than 100 studies, project modifications, and miscellaneous projects were authorized. Congress placed a cap on the amount that can be appropriated in any one year for projects authorized by the Act, rising from $1.4 billion in 1987 to $1.8 billion in 1991. In order to prevent cost overruns, the authorized cost of a project must be the maximum cost except for (1) overruns of no more than 20% of the authorized project cost where the Corps makes modifications that do not materially alter the scope or function of the project; or (2) additional studies, modifications, or actions, including mitigation and other environmental actions, that are required by law.

The central issue resolved by the Water Resources Development Act of 1986 was: Who will pay for Corps water resources projects? The following cost-sharing requirements apply to all Corps water resources projects that were not under construction as of May 1, 1986. They do not apply to Bureau of Reclamation projects or to Corps projects on most of the main stem of the Mississippi River.

1. *Harbors.* Nonfederal interests must pay, during the construction period, between 10% and 50% of the project cost, depending on harbor depth. An additional 10%, with interest, must be paid within 30 years. Nonfederal interests must also provide lands, easements, rights-of-way, most relocations, and spoil disposal areas (hereafter "ancillary costs"); but these costs are credited to the percentage cost-share. The federal share of harbor operation and maintenance (O&M) is 100%, except for deep-draft harbors, where nonfederal interests pay 50% of the costs in excess of those for a 45-foot deep harbor. A Harbor Maintenance Trust Fund is established, financed by a Harbor Maintenance Tax of 0.04% of the value of commercial cargo moving through U.S. ports. Harbor Maintenance Trust Fund monies will be earmarked to pay 40% of harbor maintenance dredging costs, which would otherwise be paid entirely by the federal government. Nonfederal interests may, on their own, plan and construct harbor projects if they comply with federal regulations. Under certain conditions, nonfederal interests constructing harbor projects may be reimbursed for the appropriate federal share. A nonfederal interest may impose port or harbor dues (in the form of tonnage duties or fees) in order to (1) finance the nonfederal share of construction and O&M costs of a completed harbor project; and (2) provide emergency response services in the harbor.

2. *Inland Waterways.* 50% of all construction costs, including ancillary

costs, must be paid from the Inland Waterways Trust Fund, a special fund financed by an inland waterways fuel tax increasing progressively from 10 cents per gallon until 1990 to 20 cents per gallon after 1994. The federal share of O&M costs is 100%.

3. *Flood Control.* For structural flood control, there is a minimum cost-share of 25% and a maximum of 50%, allocated as follows:

   (1) 5% of the costs of each project assigned to flood control must, in all cases, be paid in cash during construction.

   (2) Ancillary costs must be provided by nonfederal interests, except to the extent that they exceed 50% of project costs when added to the cash required under (1).

   (3) If the amounts contributed under (1) and (2) are less than 25%, the nonfederal interests must contribute an additional amount in cash during construction in order to reach the 25% minimum.

Nonfederal interests may pay any portion of their share exceeding 30% of project costs over 15 years (or an agreed-upon shorter period), with interest.

The cost-sharing requirement for nonstructural projects (e.g., flood-proofing, land-use regulation, land acquisition, and relocation) is 25%, with no cash required during construction. Nonfederal interests must pay all ancillary costs, even if these exceed 25% of the project costs assigned to flood control.

4. *Other Project Purposes (nonfederal share).*

   (a) Hydroelectric Power—100%;

   (b) Municipal and Industrial Water Supply—100%;

   (c) Agricultural Water Supply—35%;

   (d) Recreation—50%;

   (e) Hurricane and Storm Damage Reduction—35%;

   (f) Aquatic Plant Control—50%;

   (g) Beach Erosion Control, Water Quality Enhancement, and Fish and Wildlife Mitigation—assigned to other project purposes and shared in the same percentages as apply to those purposes, except that all costs assigned to benefit privately-owned shores (where those shores are not open to the public) or to prevent losses to private lands are to be 100% nonfederal. The federal government will bear all costs assigned to the protection of federally-owned shores.

5. *Fish and Wildlife Mitigation and Enhancement.* Mitigation must be undertaken and mitigation lands acquired either before construction or concurrent with acquisition of project lands, except that physical construction for mitigation may take place during project construction. Mitigation costs may not exceed $7.5 million per

project or 10% of project costs, whichever is greater. For mitigation, costs are to be shared in accordance with the project purposes
for which mitigation is required. Each proposal for authorization
submitted to Congress must include a mitigation plan or a finding
that mitigation is unnecessary because the project will have
"negligible adverse impact on fish and wildlife." Fish and wildlife enhancement measures that provide national benefits (where
benefits accrue to migratory or anadromous fish, or threatened or
endangered species, or occur on national wildlife refuge lands) are
a federal cost. Other enhancement is to be cost-shared on the basis of 25% nonfederal share. The Act establishes a revolving Environmental Protection and Mitigation Fund, capitalized at $35
million, for undertaking mitigation measures or acquiring mitigation lands in advance of project construction. The Fund must be
reimbursed out of the first appropriations made for construction.

6. *Lake Restoration and Streambank Erosion Control Programs.* With regard
   to these new programs, the nonfederal beneficiaries must pay 25%
   of project costs, pay all ancillary costs (credited against the nonfederal share), and operate and maintain the project after completion by the Corps.

7. *Credits to Nonfederal Interests.* Borrowing a page from Bureau of
   Reclamation regulations, Congress authorized the Corps to consider ability-to-pay in negotiating cost-sharing agreements for flood
   control or agricultural water supply. Moreover, the Corps may
   decide that work undertaken by the nonfederal sponsor in the five
   years preceding enactment is compatible with an authorized flood
   control project. In that case, the benefits and costs of the work will
   be counted toward the benefits and costs of the authorized flood
   control project, and the cost of such work can be credited to the
   nonfederal share of project costs. The nonfederal interests must
   still pay at least 5% of the project cost in cash during construction.

SOIL CONSERVATION SERVICE

The Soil Conservation Service (SCS) is an agency within the U.S. Department of Agriculture with a variety of responsibilities for soil and water conservation. For our purposes, SCS administers the "small watersheds"
program under the Watershed Protection and Flood Prevention Act of 1954,
frequently referred to as Public Law 566. This program integrates nonstructural and structural approaches to the facilitation of drainage and prevention of flooding and soil erosion. "These projects combine soil and water
conservation on the land with control and use of runoff by means of
upstream dams and other structures. Land treatment to increase vegetation,

improve soil condition, and retard surface runoff precedes structural work on the streams.''[12] The goal of the small watersheds program is watershed protection,'' the protection and development of upstream drainage areas of small tributaries, as contrasted with major projects for flood control on main streams of the type undertaken by the Army Corps of Engineers.''[12]

Unlike the Corps construction programs, the SCS does not itself carry out small watersheds projects. Local organizations such as soil conservation districts, flood control districts, irrigation districts, counties, and municipalities sponsor the projects, arrange for their activation, and supervise their operation and maintenance. The link between SCS and local sponsoring organizations is the ''state conservationist,'' an SCS employee who works directly with state and local governments and special-purpose districts. ''Thus, federal, state, and district personnel cooperate in an unusual though apparently workable arrangement.''[13] These contacts are especially important because SCS encourages the multipurpose projects that are generally favored by states. States must be provided with an opportunity to review and approve or disapprove applications by local organizations, recommend priorities for assistance within the state, review and comment on work plans, and assist in financing costs assigned to local organizations for installing and maintaining works of improvement.

The Soil Conservation Service provides three kinds of assistance to local organizations in planning and carrying out the structural components of small watershed projects: (1) technical assistance for planning, installing, operating, and maintaining works of improvement; (2) financial assistance to defray all of the cost of structural measures for flood prevention and part of the cost of structures for irrigation and drainage and for fish and wildlife development and recreation; and (3) long-term credit to help local interests finance their share of the structural costs, including the full cost of structures for municipal or industrial water supply.

In contrast to the Corps programs, SCS stresses land treatment and management practices in smaller watersheds. Land treatment measures are generally applied by private landowners with technical and financial assistance provided by the local organization and SCS. The small watersheds program, like the Corps programs, requires a favorable cost-benefit ratio for projects, but financial and technical assistance cannot be provided if the primary monetary benefits inure to the landowners by way of additional production without significant public benefits. Moreover, each project must contain benefits directly related to agriculture that amount to at least 20% of total project benefits. In order to qualify for SCS watershed protection assistance, land treatment measures must appreciably reduce erosion and sedimentation, decrease flooding, or produce drainage benefits in combination with structural techniques. If SCS is requested to finance a structure with storage capacity, over 50% of the relevant landowners in the drainage area must agree to implement recommended soil conservation measures.

By statute, small watershed projects are limited in size. The watershed area, including subwatersheds, may not exceed 250,000 acres. No structure may be larger than 25,000 acre-feet total capacity, nor may any structure have a floodwater detention capacity exceeding 12,500 acre-feet. Plans involving an estimated federal contribution to construction costs of more than $250,000, or including any single structure providing more than 2,500 acre-feet total capacity must be approved by the Public Works Committees of both the Senate and House of Representatives. Smaller projects may be administratively approved. Where there is overlap between the statutory authority of SCS and the Corps, the terms of a cooperative agreement determine which agency will prevail. In general, the Corps will provide flood protection for downstream agricultural floodplains and for urban areas with major flood problems.[12] There has been rivalry between the agencies because the SCS contends that upstream watershed protection will partially obviate large flood retention dams downstream.

SCS has not been the object of as many environmental lawsuits as the Corps. "Until the last few years of the [1960s], relationships between SCS and the organized conservation movement were excellent. The big dams, long aqueducts, and canals of the Corps and the Bureau of Reclamation were under heavy environmentalist attack, but the land treatment measures and small impoundments of P.L. 566 projects were not."[14] But beginning in about 1968, SCS "channelization" policies have caused litigation.

TENNESSEE VALLEY AUTHORITY

The Tennessee Valley Authority (TVA), a wholly owned federal government corporation, was created by the Tennessee Valley Authority Act of 1933. Its affairs are managed by a three-member board, appointed by the president with the advice and consent of the Senate for staggered nine-year terms. Originally created to improve navigation on the Tennessee River and to control floods in the Tennessee and Mississippi river basins, Congress has authorized TVA to become a multipurpose regional development agency.

The TVA has the authority to construct and license dams and other water resources development projects throughout the Tennessee River basin.

A series of nine multipurpose dams on the Tennessee River, with 24 other major dams on the tributaries, serve to provide a stairway of navigable lakes from the mouth of the river to Knoxville, Tenn., provide flood control on the river and downstream on the Ohio and Mississippi Rivers, and provide hydroelectric power, water supply and recreation for the entire region. Six of these dams are privately owned and water releases are directed by TVA as part of the regional system. The Corps operates the locks associated with the dams.[15]

Each major TVA dam has a hydroelectric power component; in fact, with its fossil fuel and nuclear power plants, TVA is the sole supplier of electric energy in an 800,000-square-mile region.[15]

The TVA receives authorizations and appropriations from Congress for dam construction. It also issues and sells bonds for the construction of facilities for generating and transmitting electric power. A portion of the proceeds from the sale of power is kept by the board to defray expenses of operating dams and reservoirs.

The TVA has been a prime target of environmentalist lawsuits, including the notorious Tellico Dam ("snail darter") litigation that it lost in the U.S. Supreme Court. This case is discussed in Chapter 13 under the federal Endangered Species Act.

BONNEVILLE POWER ADMINISTRATION

This federal agency markets almost all hydroelectric power produced at Corps and Bureau dams in the Pacific Northwest. Although BPA cannot construct or operate power-producing facilities, it has built and it operates the power grid through which federal power is supplied. "The role of BPA in marketing includes the scheduling of power from [federal] sources to meet its loads. This scheduling function implies a large degree of control over the operations of hydropower generators and power-storage reservoirs. BPA is therefore a major actor in the allocation of streamflows over time."[16]

Chapter 11

THE FEDERAL GOVERNMENT AS PROJECT LICENSOR

This chapter analyzes the federal government's role, under the Commerce Clause, in controlling the use of water by nonfederal development entities. Emphasis is placed on the licensing activities of the Nuclear Regulatory Commission, the Federal Energy Regulatory Commission, and the Army Corps of Engineers as they affect water resources development. Federal grant programs for construction of waste treatment plants and sewers, as well as federal Clean Water Act permitting, are covered in Part IV.

NUCLEAR REGULATORY COMMISSION

The Nuclear Regulatory Commission (NRC) was created as an independent agency in 1974. It was delegated responsibilities formerly carried out by the Atomic Energy Commission under the Atomic Energy Act of 1954. The Commission's major importance in the field of water resources development stems from its authority to license the construction and operation of nuclear power plants. Nuclear plants must be sited on watercourses or saltwater environments because of their extensive need for cooling water. Although the NRC exclusively regulates the safety-related aspects of nuclear plant construction and operation, states control the determinations of power plant need, type, and siting. Another area of NRC authority that can affect water resources is its search for appropriate sites in which to store highly radioactive spent fuel from nuclear reactors.

Neither the Environmental Protection Agency (EPA) nor a state is empowered to regulate radioactive discharges from nuclear plants. States and the EPA can, however, set effluent limitations and water quality standards for the thermal component of nuclear plant discharges. But if these regulations on heat discharges significantly disrupt a plant's operation, the state or EPA will probably be compelled to defer to the NRC. Because of NRC preemption of safety factors in plant operation and the close relationship between a plant's operation and its thermal effluent, courts generally deny recovery on traditional legal theories (e.g., nuisance, negligence) to neighboring riparian landowners or states, suing as public trustees of wildlife, that suffer injury from NRC-approved thermal discharges of nuclear power plants. However, since power companies possess the right of eminent domain, local property owners may be able to win condemnation awards if the damage has been severe enough.

FEDERAL ENERGY REGULATORY COMMISSION

The Federal Energy Regulatory Commission (FERC) was created in 1977 within the U.S. Department of Energy. Many of the functions previously performed by the Federal Power Commission (FPC), now defunct, have been assumed by the FERC. The FERC programs that affect water resources development are its authority to license nonfederal hydroelectric projects and its responsibility to approve the location of natural gas pipelines. The FERC licensing process applies to a wide variety of developers including investor-owned utilities, private entrepreneurs, industries, municipalities, state power authorities, irrigation districts, and electrical cooperatives.[17] Federal project development agencies do not require FERC licenses or permits. The FERC is authorized to issue preliminary permits for site study, minor project licenses, major project licenses at existing sites, major project licenses at new sites, and small hydroelectric power exemptions.

Prior to 1920, Congress granted the right to construct, operate, and maintain hydroelectric projects by legislative action in particular cases.[15] This separate, piecemeal water resources development was considered by conservationists to be inefficient and dangerous to resource conservation. "The Federal Water Power Act of 1920, part I of the present Federal Power Act, was seen as a great step toward the regulated and comprehensive development of the Nation's water resources and enjoyed the support of the conservationist movement."[5]

> Although the title of the Federal Water Power Act indicated that its primary thrust was the control and development of hydroelectric power, it is clear that

the act required that such control and development be part of a comprehensive plan. The purpose of the act would be attained only if the Commission considered the impact of the project in the full spectrum of the public interests, rather than considering only some limited or partial aspect of commerce and the public interests.[18]

Thus, the Federal Power Act (FPA) provides that licenses be issued on condition that the project "be best adapted to a comprehensive plan for improving or developing a waterway or waterways for the use or benefit of interestate or foreign commerce, for the improvement and utilization of waterpower development, and for other beneficial public uses, including recreational purposes." The concept of comprehensive watershed planning is the backbone of the FPA. A comprehensive plan must be developed before a license (and in some cases before a preliminary permit) may be issued.

One of the most famous of all water resources-related environmental cases examined the FPC's duty to evaluate a project comprehensively, including the alternatives to it and its adverse environmental effects. The project involved in *Scenic Hudson Preservation Conference* v. *Federal Power Comm.*[18] was Consolidated Edison's proposed pumped storage Cornwall Project at Storm King Mountain on the Hudson River. The FPC issued a license, but the second circuit overturned the agency's decision because the FPC had not actively sought out information on power alternatives and environmental impacts:

> In this case, as in many others, the Commission has claimed to be the representative of the public interest. This role does not permit it to act as an umpire blandly calling balls and strikes for adversaries appearing before it: the right of the public must receive active and affirmative protection at the hands of the Commission.

Scenic Hudson makes it clear that the FERC, after balancing the various public interests, may deny a license where power benefits are outweighed by recreational and ecological losses.

Nevertheless, environmentalists have been consistently skeptical about the FERC's sensitivity to environmental concerns. In 1986 Congress amended the FPA to require that the FERC (1) give "equal consideration" with power generation to energy conservation and environmental quality, (2) consider any federal or state comprehensive plans for a waterway, and (3) include in all licenses and license exemptions conditions to protect fish and wildlife recommended by the federal Fish and Wildlife Service or National Marine Fisheries Service and state fish and wildlife agencies; if the FERC disagrees with a recommendation, it must issue a supported finding that the condition is illegal or unnecessary.

The Commission's power to license nonfederal hydroelectric projects extends to three areas: (1) the licensing of dams, conduits, powerhouses,

transmission lines, and other project works for the development or transmission of power, whether the works be across, along, or in any waterbody
which Congress may be capable of regulating under the Commerce Clause;
(2) the licensing of hydroelectric facilities on public lands; and (3) the licensing
of projects using surplus water from federal dams. Although the FERC normally requires an applicant to satisfy all provisions of state law (e.g., state
hydropower licensing or water diversion regulations) before applying for
a FERC license, courts invariably hold that the FERC, at its discretion, may
preempt state law and assert exclusive jurisdiction over hydropower facilities. Moreover, a license applicant need not own the site. Should a license
be granted, the FPA allows a licensee to condemn the site and any water
diversion rights necessary for the project.

Powerful as it is, the FERC does not have complete control over the
hydropower licensing process. As mentioned above, conditions stipulated
by federal and state fish and wildlife agencies must be adopted by the FERC
or the agency must formally explain why not. Also, where navigation may
be affected by a hydropower structure, the Corps must approve plans and
specifications. The limitations on federal licensing discussed later in this part
apply to the FERC as well as the other federal water development and licensing agencies.

The FERC does not have the authority to construct hydroelectric facilities. But, it may deny a license if it decides that the development should
be undertaken by the United States. Congress would then have a reasonable opportunity to study the FERC recommendations and determine
whether to authorize a federal project. The FERC may condition the license
upon the applicant's installation of navigation devices, e.g., locks, booms,
sluices, or fish ladders. Minimum flows or releases may be required. If public
lands are involved, the FERC may impose conditions to protect them. Each
applicant must submit a "recreation plan" maximizing public recreation at
the site. Licensees are expected to acquire, and include within the project
boundary, enough land to assure optimum development of recreational
facilities, to develop suitable recreation on project lands and waters, and
to encourage public access.[17] Licensees may include these costs in their reimbursable rate bases or charge admission fees.

During the energy shortage of the 1970s, Congress enacted a package
of incentives for the development of small hydropower facilities. The
incentives included: (1) exemptions from licensing for relatively small facilities using existing dams, conduits (tunnels, canals, and pipelines primarily
intended for agricultural, municipal, and industrial water distribution), and
natural water features; (2) guaranteed purchase by public utilities of power
from these small hydropower facilities; (3) loan programs to stimulate project
development; and (4) tax incentives. Although the statutes appeared to limit
all these incentives to existing dams, the FERC extended them to new dams
as well.[19] The FERC's expansive interpretation stimulated a boom in small

hydropower applications, especially in California, New England, New York State, and Appalachia.

Environmental groups, states, and Indian tribes all felt threatened by "hydromania," as it was called.[20] The environmentalists criticized the FERC for neglecting comprehensive planning, failing to consider the cumulative effects of multiple hydropower projects in a single watershed, and ignoring environmental impacts, especially adverse effects on anadromous fish. States complained that the FERC was preempting state laws, facilitating private condemnation of state water rights, and paying no heed to state comprehensive plans. Indian tribes accused the FERC of disregarding their interests while granting licenses on reservations. Each of these groups turned to the courts to vindicate their claims. For example, in one case the Ninth Circuit Court of Appeals held that the FERC may not grant exemptions for projects involving new dams or impoundments.[21]

They also turned to Congress, and the 1986 amendments spoke to some of their concerns. Indian tribes were reassured that the FERC may not license a project that is inconsistent with the purpose of a reservation. Moreover, Congress substantially qualified the FERC's "new dam" interpretation. New dams will only be eligible for small hydropower incentives where a project (1) "will not have substantial adverse effects on the environment," and (2) is not located on a stretch being studied for inclusion in, or actually included in, a federal or state wild and scenic river system. Projects with already-filed license applications are not required to meet these new tests, but licenses will only be granted subject to conditions recommended by fish and wildlife agencies.

The 1986 amendments, however, did not weaken the FERC's capacity to exercise exclusive jurisdiction, vis-a-vis states, over hydropower licensing; nor did they inhibit the FERC from authorizing condemnation of state-perfected diversion rights. Future hydropower litigation and legislation will probably center around these federal-state relations in addition to the meaning of "equal consideration" for nonenergy considerations, how much consideration the FERC must give to federal and state comprehensive plans, the extent of its own obligation to plan comprehensively, and the environmental criteria and grandfather exemptions relating to new dams.

ARMY CORPS OF ENGINEERS

The Corps has authority under various statutes to regulate water resource development in tidal and nontidal waters. Most of this authority arose from "the Rivers and Harbors Acts passed by Congress for the last one-hundred-odd years. These provisions have been loosely codified in Title 33 of the *United States Code*. They are characterized by imprecision, inconsistency, and incongruity."[22] Thus, there has been a good deal of litigation in this area.

Fortunately, authoritative judicial interpretations of Corps enabling acts are incorporated into current Corps regulations.[23] These regulations are the best source of guidance as to Corps regulatory policies and practices. However, administrative regulations are not sacrosanct. Courts can overturn them for any number of reasons—for example, exceeding statutory authority, being "arbitrary and capricious," or violating procedural or constitutional safeguards.

Section 9 of the Rivers and Harbors Act of 1899 prohibits the construction of any dike or dam that completely spans a navigable waterway without congressional consent and Corps approval of plans. Where the navigable portions of the waterbody lie wholly within the limits of a single state, the dam or dike may be built under authority of the legislature of the state if the location and plans are approved by the Corps. Corps approval for these projects is in the form of a permit. Bridges and causeways are regulated by the U.S. Department of Transportation (Coast Guard), not the Corps.

Section 10 of the Rivers and Harbors Act of 1899 prohibits the unauthorized construction in or alteration of any navigable water of the United States. It requires a Corps permit for "the construction of any structure in or over any navigable water . . . , the excavation from or depositing of material . . . , or the accomplishment of any other work affecting the course, location, condition or capacity of [navigable] waters." The Corps also has jurisdiction over artificial islands, installations, and other devices located on the outer continental shelf, generally beyond the three-mile limit.

Central to understanding the legal impact of section 10 and other Corps programs is the Corps' definition of "navigable waters of the United States." This concept limits the Corps' jurisdiction and designates the area of federal navigation servitude, where the Corps is not required to pay compensation if its activities (e.g., construction projects or permit denials) significantly reduce land values. Corps regulations define navigable waters as

> those waters that are subject to the ebb and flow of the tide and/or are presently used, or have been used in the past, or may be susceptible for use to transport interstate or foreign commerce. A determination of navigability, once made, applies laterally over the entire surface of the waterbody, and is not extinguished by later actions or events which impede or destroy navigable capacity.

Tidal waters are presumed to be navigable. The navigability of nontidal waters depends on their suitability for commerce. "Commerce" may be shown by historical use of canoes for commercial purposes, regardless of portages around obstructions, or by commercial transportation of logs. Where a waterbody is located entirely within a state, it is considered navigable "when it physically connects with a generally acknowledged avenue of interstate commerce, such as the ocean or one of the Great Lakes." As for interstate waterbodies, "Where a waterbody extends through one or more states, but substantial portions, which are capable of bearing interstate commerce, are located in only one of the states, the entirety of the waterway up to the head (upper limit) of navigation is subject to Federal jurisdiction." To review, a navigable water of the United States must (1) be, or have been; (2) used or susceptible of use; (3) in the customary modes of trade and travel by water; (4) as a highway for interstate commerce.

Artificial channels (canals) are navigable waters if they are used, or are capable of being used, for commerce. This is true even if the canal has been privately developed or passes through private property. "Unplugged canals," open on one or both ends to navigable waters or tide-flowed, are generally navigable, while "plugged" canals are frequently not. Susceptibility for use in commerce depends upon feasibility of commerce after the construction of reasonable improvements. "The improvements need not exist, be planned or even authorized; it is enough that potentially they could be made." Once a waterbody has been navigable, or could have been navigable, it is legally navigable forever, despite its inability to presently be used for navigation. Thus, an unplugged canal which has been plugged remains navigable in a legal sense.

Federal jurisdiction and the navigation servitude over rivers and lakes include all lands and waters below the ordinary high-water mark of navigable waters. Ocean and coastal waters are presumed to be navigable up to "the line on the shore reached by the plane of the mean (average) high water." This boundary is referred to as the "ebb and flow" line, and divides "section 10 waters" from "fast" land or land above the mean high-water mark. "Where precise determination of the actual location of the line becomes necessary, it must be established by survey with reference to the available tidal datum, preferably averaged over a period of 18.6 years." Coastal marshlands are legally navigable only if subject to inundation by mean high waters.

Now that what the Corps means by "navigable waters" is understood, it is possible to appreciate the importance and controversiality of section 404 of the Clean Water Act, which requires a Corps permit for discharge of dredged or fill material into "waters of the United States." Except for dredging unaccompanied by filling, section 404 covers the same activities as section 10. But the scope of section 404 is much broader than "navigable waters." Courts have given "waters of the United States" the broadest possible interpretation consistent with the U.S. Constitution; intrastate streams, freshwater wetlands, drainage ditches, mosquito canals, and intermittent streams are within the reach of section 404. But the Corps, which considers itself a navigation enhancement and downstream flood protection agency, has resisted any extension of its regulatory authority beyond traditionally navigable waters. There has been a continuing battle over this between environmentalists, on the one hand, who desire the Corps to regulate filling over a wide geographical spectrum, and the Corps, which feels uncomfortable regulating anything but navigable waters. Section 404 and another Corps regulatory provision—section 103 of the Marine Protection, Research and Sanctuaries Act of 1972 that governs Corps permits for ocean dumping of dredged spoil—are described more thoroughly in Parts III and IV.

Where navigability has been established, the Corps is more receptive to considering factors in addition to navigation. As a result of several federal statutes and court decisions, the Corps conducts a "public interest review" as part of its section 10 permitting process. This is a general balancing of project benefits and detriments, including impacts on conservation, economics, aesthetics, general environmental concerns, wetlands, cultural values, fish and wildlife values, flood hazards, floodplain values, land use, navigation, shore erosion and accretion, recreation, water supply and conservation, water quality, energy needs, safety, food and fiber production, mineral needs, and property ownership.[23] Understandably, the emphasis is on navigation, but the Corps has denied section 10 permits "in the public interest" on solely ecological grounds.

Corps permits are of two kinds, individual and general. A general permit may be either regional or nationwide. No separate application or authorization is required for a general permit. It is an authorization on a regional or national basis for a category of activities when "those activities are substantially similar in nature and cause only minimal individual and cumulative environmental impacts." For example, general section 10 permits may be issued for: (1) structures in previously authorized artificial canals "within principally residential developments"; (2) repair or replacement of previously authorized, "currently serviceable" structures; (3) small bank stabilization projects; and (4) dredging of less than 10 cubic yards. General permits last for five years (subject to renewal) and incorporate certain conditions, such as protection of public water supply intakes, protection of fish and wildlife, and the application of certain management practices. Division Engineers have the discretion to override general permits and require individual permits. Or, general permits may require case-by-case reporting and acknowledgment systems.

The Corps has a number of enforcement options where work has been done without a section 10 permit or where an existing permit has been violated. If the violation is innocuous, the Division Engineer may accept an application for an "after-the-fact" permit. If the unauthorized activity is unacceptable, he may issue a "stop work order," a "restoration order," or both. In issuing a restoration order, the Division Engineer must consider the degree and kind of environmental disturbance caused by the violation as well as the practicality of restoration. Existing permits may be revoked or modified for failure to obey permit conditions. In serious cases, the Division Engineer may refer the matter to the local U.S. Attorney for civil or criminal action. Only the Corps may sue to enforce section 10; an individual citizen or citizen's group has no standing under this statute.

Chapter 12

COMPREHENSIVE WATER RESOURCES PLANNING AND RESEARCH

WATER RESOURCES PLANNING

Water resources planning takes place whenever a decision is made about water resources development or protection:

> Water resources planning is carried out at every level of government as well as by private industry. Planning is not decisionmaking but is the prelude to informed decisionmaking. A considerable portion of the Nation's water planning is done in urban areas by city water and sewer departments, sanitary districts, and drainage and flood control districts. In rural areas, local water planning is being done by such agencies as soil conservation districts, watershed districts, and irrigation associations. Some States now have statewide water plans. In many States there are intrastate basin planning organizations, taking many forms, from irrigation districts to river basin authorities.[24]

Interstate agencies and the federal government also perform water resources planning. At the federal level, planning evolved as a preliminary to water resources development activities by federal agencies.

The major deficiency of American water resources planning has been its fragmentation: water planning has not been adequately integrated with land use planning; planning for river basins has neglected the water-related needs of metropolitan areas; environmental planning, or water quality planning, has been separated from water resources development planning; and federal agency plans have been project-specific, mission-specific, and uncoordinated with water resources plans of other federal agencies and non-

97

federal organizations.[24] Consequently, comprehensive water resources planning has become the goal of many water resources professionals. Comprehensive water resources planning has two major aspects: (1) better coordination of the different types and levels of planning, both among Federal agencies and between Federal and non-Federal planners (i.e., regional or river basin planning); and (2) broadening the traditionally narrow developmental objectives of federal water resources planning to include multiobjective planning.

The Water Resources Planning Act (WRPA) of 1965 was enacted in order to promote comprehensive water resources planning. In 1981, President Reagan nullified the WRPA by dissolving the Water Resources Council and the river basin commissions. However, familiarity with the WRPA's provisions is essential for three reasons: (1) Congress has not repudiated the WRPA, and its institutional system could be reinstated by a more sympathetic administration; (2) there is a clear need for comprehensive water resources planning; and (3) the WRPA experience will serve as valuable data for future institutional development.

In the WRPA, Congress attempted to "encourage the conservation, development, and utilization of water and related land resources . . . on a comprehensive and coordinated basis by the Federal Government, States, localities, and private enterprise." The Act established a Water Resources Council, composed of the secretaries of interior, agriculture, army, commerce, and housing and urban development, the administrator of the Environmental Protection Agency, and the chairman of the Federal Energy Regulatory Commission. The council's chairman, as designated by the president, was the secretary of the interior.

The role of the Water Resources Council was fivefold: (1) preparing the national assessment of water supply and demand; (2) developing principles, standards, and procedures for project formulation and evaluation; (3) establishing, and maintaining liaison with, river basin commissions, and formulating principles and standards for the preparation of comprehensive regional or river basin plans; (4) making grants to states for water planning; and (5) reviewing river basin plans. The Water Resources Council was not given a project review function by the WRPA. But the "principles and standards" for project evaluation were recognized as potentially having a crucial hortatory effect on federal development agencies.

The principles and standards were developed from the Flood Control Act of 1936 that approved federal investment and participation by the Corps in flood control projects "if the benefits to whomsoever they may accrue are in excess of the estimated costs, and if the lives and social security of people are not otherwise adversely affected." This "benefit-cost analysis" was extended to other federal development agencies, making project justification depend upon a favorable ratio of benefits to costs. The principles and standards were intended to clarify and rationalize benefit-cost analysis, and to resolve such contentious issues as the objectives of federal water resources development, the discount rate for future benefits, and the treatment of project alternatives.

After years of controversy, the principles and standards were issued in 1973. They defined the objectives of water and land resource planning as national economic development and environmental quality, with regional development and social well-being as secondary objectives.[25] The principles and standards established a detailed planning process, with instructions for the formulation of alternative plans and analysis of trade-offs among alternatives, including nonstructural alternatives. In practice, these principles and standards had little effect on agency planning, partly because the Water Resources Council could not enforce them through project reviews. When President Carter announced his water resources policy, it included reform of the Water Resources Council and its principles and standards. Carter ordered the Council to revise the principles and standards to emphasize water conservation and nonstructural alternatives. Then, he issued an executive order requiring an independent water project review by the Water Resources Council. However, the Water Resources Council was unable to implement this order because of congressional opposition. President Reagan revoked the executive order, rescinded the principles and standards, and finally abolished the Water Resources Council. New "principles and guidelines" became effective in 1983. They are not legally binding on federal planners.

Under the WRPA, one of the pivotal responsibilities of the Water Resources Council was to supervise the work of the new river basin commissions and make recommendations to the president for integrating river basin plans with federal water resources development. A river basin commission was to be established by presidential executive order on request of the Water Resources Council or a state. The Council and at least half of the basin states were required to consent to a commission's establishment. River basin commissions were to (1) coordinate federal, state, interstate, and local development plans for the basin, (2) prepare and update a "comprehensive coordinated joint plan," including evaluations of and alternatives to individual projects, (3) set priorities for data collection, and (4) undertake studies necessary to prepare the plan. River basin commissions possessed no regulatory or construction authority; they were intended as consensus-building and advisory agencies. In all, only seven river basin commissions were established under the WRPA, and their plans were of uneven quality.

The structure of river basin commissions inhibited their effectiveness.[24] The chairman of the commission was appointed by the president, but could not be a member of a federal agency. Except for the chairman, all commission members represented, and were paid by, other organizations. Each federal agency with a substantial interest in the river basin, each state in the basin, and each appropriate interstate or international agency was entitled to membership. But nothing could be accomplished without con-

238225

sensus of all members. River basin commissions also operated with uncertain budgets. Neither Congress nor the member states appropriated money to the river basin commissions. Instead, Congress appropriated money directly to the member federal agencies, and states appropriated money directly to the member state agencies. Thus, the river basin commissions lacked control over their own budgets.

Professor Ingram sees the powerlessness and structural paralysis of the river basin commissions as illustrative of a problem that is common to all regional agencies—lack of political viability:

> The expectation of the designers is that regional entities will provide a comprehensive, coordinated, regional approach to water resources. In practice, it is extremely difficult to demonstrate that the fragmented, disjointed approach which typifies and, in the view of many, plagues water policy-making has been much altered by establishment of regional water institutions. . . . (F)ailure of regional institutions to accomplish real innovation is associated with the restraints imposed by political viability. To be viable, that is, to survive, regional agencies have had to tailor their actions to build support. Support-building actions then compromise their ability to approach water and other resource problems in a comprehensive, coordinated, regional manner.[26]

Perhaps decentralized water resource planning is as endemic to our political system as decentralized housing or transportation planning.

WATER RESOURCES RESEARCH

The nation's commitment to water resources research was continued when Congress overrode President Reagan's veto and enacted the Water Resources Research Act of 1984. Water resources research in the United States centers around the cooperative relationship between the U.S. Department of the Interior, presently acting through the U.S. Geological Survey (USGS), and the 54 state water resource research institutes located at land-grant universities in each of the 50 states and Puerto Rico, the District of Columbia, Guam, and the Virgin Islands.[27]

The Water Resources Research Act of 1984 authorizes 10 million dollars per fiscal year through fiscal 1989 for matching grants to state institutes. The matching state share is one-to-one for fiscal 1985 and 1986; one and one-half nonfederal dollars for each federal dollar in fiscal 1987 and 1988; and two-to-one in fiscal 1989. This reflects the trend toward cost-sharing by recipients of federal subsidies.

State institutes, along with other educational institutions, private foundations and firms, individuals, and agencies of state or local governments, are also eligible for matching grants "for research concerning any aspect of a water resource-related problem which the Secretary may deem to be in the national interest." Twenty million dollars per fiscal year through fiscal 1989 has been authorized for these grants, with a matching share of one-to-one. The matching requirement may be waived altogether in special cases where basic research is involved.

Third, the Act authorizes six million dollars per fiscal year for grants to qualified recipients "for technology development concerning any aspect of water-related technology which the Secretary may deem to be of State, regional, and national importance, including technology associated with improvement of waters of impaired quality and the operation of test facilities." Matching requirements here are flexible.

Funding priorities should be consistent with the Act's express purposes, which are to:

1. Assure supplies of water sufficient in quantity and quality to meet the nation's expanding needs for the production of food, materials, and energy.
2. Discover practical solutions to the nation's water and water resources related problems, particularly those problems related to impaired water quality.
3. Assure the protection and enhancement of environmental and social values in connection with water resources management and utilization.
4. Promote the interest of state and local governments as well as private industry in research and the development of technology that will reclaim waste water; convert saline and other impaired waters to waters suitable for municipal, industrial, agricultural, recreational, and other beneficial uses.
5. Coordinate more effectively the nation's water resources research program.
6. Promote the development of a cadre of trained research scientists, engineers, and technicians for future water resources problems.

State institutes are specifically directed to encourage research into supply and demand for water, demineralization of saline and other impaired waters, conservation and reuse, depletion and degradation of groundwater supplies, multidisciplinary aspects of water resource management, and effective communication of research results. The quality of institute research will be monitored by the USGS and evaluated with the assistance of multidisciplinary advisory teams that make periodic site visits.

Conspicuous by its absence from the Act is a direction, found in prior statutes, to establish a national center or clearinghouse within the Department of the Interior for the acquisition, processing, and dissemination of information dealing with all areas of water resources research, technology development, and demonstration. There is an indisputable need for a water resources research information center, but its structure and location is a contentious issue within the water resources research community. Some water resources professionals favor leaving this function with the USGS; others prefer a nongovernmental clearinghouse run by a consortium of universities or a private organization.

Chapter 13
LIMITATIONS ON FEDERAL DEVELOPMENT AND LICENSING

The environmental movement has had a profound influence on water resources development. In evaluating a proposed project, environmental protection has now achieved legal parity with efficiency and equity. Federal development decisions are no longer made by the "iron triangle" of local development interests and their congressional representatives, friendly congressional committees, and the federal development or licensing agency. Environmental considerations are no longer subordinated to economic benefits or subsumed within a monetized benefit-cost analysis. The "pork barrel" is no longer as enticing or accessible as it once was, due in good measure to the enactment of environmental protection requirements. Indeed, the value that our society places on environmental protection has served to "democratize" the federal water resources development process—to open it up to interest groups of all persuasions that had not been adequately represented in the past. This trend can only be enhanced by mandatory cost-sharing.

Three federal statutes—the National Environmental Policy Act (NEPA), the Endangered Species Act (ESA), and the Wild and Scenic Rivers Act (WSRA)—have primarily introduced environmental protection into federal water resources decisionmaking. These three major actors have been supported by a cast of lesser statutes and executive orders.

MAJOR ENVIRONMENTAL PROTECTION STATUTES

The National Environmental Policy Act revolutionized federal water resources development, licensing, planning, and research. It requires that environmental impacts of a proposed action and its alternatives be identified early in the planning process, and that these factors be brought to the attention of concerned citizens, interested governmental officials, and if necessary, the courts, before a project can be initiated. Since 1970, when the NEPA was signed into law, literally hundreds of water resources projects have been withdrawn or modified because they could not stand the light of the NEPA's day.

Section 102(c) of the NEPA requires each federal agency that proposes a major action having a significant effect on the human environment to prepare and circulate, in draft and final form, an environmental impact statement (EIS) disclosing the environmental impacts of the proposed action, reasonable alternatives and their impacts, long- and short-term effects, cumulative effects, irretrievable commitments of resources, and the agency's balance among environmental protection and economic development, national security, or other factors. Even where the proposing agency believes that a proposed action will not have a significant environmental impact, it must nevertheless prepare and circulate a mini-EIS, called an "environmental assessment," accompanied by a finding of no significant impact (FONSI). Federal actions covered by the NEPA are not only construction projects but also licenses, leases, contracts, permits, loans, and research projects which may have significant effects on the environment. The EIS may be programmatic, regional, or site-specific; various levels of EISs may be "tiered." When an agency does not comply with the NEPA's requirements, a court may, if it serves the public interest, enjoin the proposed action until an adequate EIS has been completed and disseminated.

Before the NEPA, "mission-oriented agencies" such as the Corps and the Bureau contended that they had no legal authority to consider environmental protection even if they wanted to do so. The NEPA mandates that all federal agencies consider the environment "to the fullest extent possible"; that is, unless there is an obvious inconsistency between the NEPA and an agency's enabling legislation. Agencies must strictly comply with the NEPA; environmental concerns must be integrated into the agency's planning process at the earliest possible time. When a project is first being seriously considered, the proposing agency must hold a "scoping meeting" and invite representatives of interested groups, including groups that might be critical of the project. Subsequent project planning must also include environmental planning.

The Corps planning process, described in Chapter 10, is an example of how one federal agency has factored the environment into its development decisions. Environmental impacts must be identified in the Reconnaissance Report, which is prepared with public participation. At the Feasibility Report stage, public participation is once again encouraged to further explore the environmental effects of the proposed project and reasonable alternatives, as well as potential mitigation measures. During this phase a Draft Environmental Impact Statement (DEIS) is prepared and circulated for comments to federal agencies with relevant environmental expertise (e.g., the EPA and the Fish and Wildlife Service), state and local officials, and concerned citizens' groups. Responsible comments must be included in the Final Environmental Impact Statement (FEIS) that becomes an integral part of the Corps review process and is ultimately transmitted to Congress before it authorizes funds for construction.

The NEPA does not mandate particular results; it does not compel a federal agency to adopt the least environmentally damaging alternative, although the agency must explain why it was not chosen. Nor does the NEPA elevate the environment to a position superior to other policy concerns, such as economic development or national security. Courts frequently refer to the NEPA as an "environmental full disclosure law," obliging an agency to take a "hard look" at the environmental consequences of its actions. A complete EIS, subject, of course, to the NEPA's "rule of reason," is conclusive evidence that the proposing agency has given "good faith consideration" to the environment. Once the EIS is complete (i.e., the agency makes full disclosure) the political process, not the courts, determines the outcome, unless the proposal is so outrageous that it is "arbitrary and capricious." In most instances, the NEPA does guarantee that other federal agencies having expertise in the area, state and local officials, and the general public will have an opportunity to formally comment on a proposed federal action before it is undertaken, and that Congress will have access to relevant information on which to make authorizations and appropriations.

The NEPA's conclusive presumption that an adequate EIS indicates good faith consideration of the environment is subject to abuse. All too often the EIS becomes a massive, unreadably technical rationalization of an action chosen on political grounds. The Council on Environmental Quality (CEQ), also established by the NEPA, has issued regulations to curtail this superabundance of NEPA riches.[28] These regulations are binding on federal agencies and enforceable in court. They are also the lay person's most reliable source of information on current NEPA law.

Approximately 30 states have enacted "little NEPAs" applicable to state and sometimes local projects and approvals.

The Endangered Species Act has four major environmental protection mechanisms: (1) listing of endangered species, (2) agency consultation and protection duties, (3) a mandatory agency responsibility to conserve endangered species, and (4) a prohibition against "takings" that is independent of federal agency action. The ESA cannot be invoked unless a species and its "critical habitat" are listed by the Department of the Interior as "endangered" or "threatened." (Threatened species receive less protection than endangered species.) Economics cannot be considered in listing a species, but is relevant to delineating its critical habitat.

Once listing has been completed, any federal agency contemplating construction or licensing in an area that might contain an endangered species must ask the Fish and Wildlife Service (FWS) whether the species is present or not. If the FWS advises an agency that such species is likely to be present, the agency must prepare and submit a "biological assessment" to the FWS. The biological assessment identifies any listed species or critical habitat in the area of the proposed project and describes the effects that the project may have on the species and habitat. If the assessment identifies potential adverse impacts, the FWS must prepare a "biological opinion" for the proposing agency. The biological opinion determines whether or not the project "is likely to jeopardize the continued existence of listed species or result in the destruction or adverse modification of critical habitat." The biological opinion may also recommend project modifications or conservation measures to mitigate any adverse impacts.

Almost all conflicts between proposed development and endangered species are resolved at this consultation stage by the incorporation of mitigation measures into project planning. But in the unusual case of an irreconcilable conflict, the agency is forbidden to undertake or license any project that is likely to jeopardize an endangered species or its critical habitat. The only litigated example of an unreconcilable conflict was *TVA* v. *Hill*,[29] the famous snail darter case, where the U.S. Supreme Court enjoined the nearly completed Tellico Dam because the impoundment would have resulted in total destruction of the snail darter's habitat. (Later, in a sort of rump session with few members present, Congress declared the Tellico Dam to be in compliance with the ESA, allowing the impoundment to be closed.)

After the *Hill* case, Congress amended the ESA to create an exemption from the statute's previously absolute bar. If an irreconcilable conflict still persists after analysis of alternatives and mitigation planning, an Endangered Species Committee may grant an exemption. Composed of federal officials and a representative of the affected state, the committee can only grant an exemption if (1) there is no "reasonable and prudent" alternative to the agency action; (2) the benefits "clearly outweigh" the benefits of conservation alternatives; and (3) the agency action is of regional or national significance. No exemption has yet been granted, and only one has been sought.

In addition to its prohibition on agency action that would be likely to jeopardize the continued existence of an endangered species or destroy its critical habitat, the ESA imposes on federal agencies a mandatory responsibility to conserve endangered species: "All Federal departments and agencies shall seek to conserve endangered species and threatened species and shall utilize their authorities in furtherance of [these] purposes. . . ." In *Carson–Truckee Water Conservancy District* v. *Clark*,[30] the Ninth Circuit Court of Appeals relied on this language in upholding a Bureau of Reclamation decision to use federal project water for endangered species protection rather than municipal and industrial water supply where there was insufficient water for both purposes.

The final provision of the ESA that might affect water resources development is its prohibition on "taking" an endangered species. "Taking" is broadly defined by the statute as including to "harass, harm, pursue, hunt, wound, . . . or attempt to engage in any such conduct." Theoretically, whenever any endangered species is adversely affected by a project there is a violation of this section. In order to forestall litigation, Congress amended the ESA to establish an exemption from the ESA's "taking" clause. The Secretary of the Interior may grant a permit for an activity that might otherwise constitute a "taking" of an endangered species if he finds that: (1) the taking will be incidental; (2) the applicant will, to the maximum extent practicable, minimize and mitigate the impacts of such taking; (3) the applicant will ensure that adequate funding for the plan will be provided; and (4) the taking will not appreciably reduce the likelihood of the survival and recovery of the species in the wild.

The Wild and Scenic Rivers Act establishes a national Wild and Scenic Rivers System and defines criteria for eligibility under each of the three classifications of the law: wild, scenic, and recreational rivers. This process is examined in detail in Part III. The Act provides that "no department or agency of the United States shall assist by loan, grant, license, or otherwise in the construction of any water resources project that would have a direct and adverse effect on the values for which such river was established, as determined by the Secretary charged with its administration." As with the ESA, this prohibition is preceded by mandatory consultation between the proposing agency and the appropriate federal resource protection agency. The WSRA subjects FERC-licensed projects to an even stricter standard. The FERC is prohibited from licensing water projects "on, or directly affecting" protected rivers. Thus, it would be illegal for the FERC to license a small hydropower facility on a tributary of a scenic river where the project has any effect on water flow in the main stream, regardless of any other impacts on scenic values. The WSRA limitations outlined above are applicable not only to rivers actually included in the Wild and Scenic River System, but also to rivers being studied for inclusion in the System.

OTHER LIMITING STATUTES AND ORDERS

The National Historic Preservation Act of 1966 established the National Register of Historic Places. It also requires federal agencies to consult with the Advisory Council on Historic Preservation whenever federal projects could have adverse impacts on these historic or archeological sites. In 1976 the Act was amended to extend its protection to properties eligible for listing on the National Register of Historic Places. By regulation, the proposing federal agency must consult a state historic preservation officer (SHPO) when determining how its activities will affect historic or archeological sites. The procedures also require that the SHPO, along with the advisory council and the proposing agency, reach written agreement in certain cases on how to mitigate adverse effects expected from a federal project.

Under the Reservoir Salvage Act, a federal agency, before beginning construction or before issuing a license for the construction of a dam, must give notice to the Secretary of the Interior of any historic or archeological data that have been found on the site during preliminary investigation. If the Secretary determines that the data are significant and in danger of being destroyed, he must conduct a survey of the project area and recover the data that he considers irreplaceable. Up to 1% of appropriated project funds may be used for the survey and recovery operations.

The Fish and Wildlife Coordination Act requires federal agencies sponsoring or issuing permits for water projects to consult with the FWS "with a view to conservation of wildlife resources," and also requires mitigation measures to minimize adverse impacts. As with the ESA, WSRA, and the National Historic Preservation Act, the Fish and Wildlife Coordination Act's mandatory consultation procedures have been integrated with agency preparation of environmental impact statements.

The Coastal Zone Management Act provides for federal financial assistance to coastal state governments for the development and implementation of coastal zone management plans. These plans have, as their primary function, land use management for the coastal zone to assure the orderly and environmentally sound development of these ecologically sensitive areas. Under the Act, a federal or federally assisted or licensed project is required to be, "to the maximum extent practicable," consistent with an approved coastal zone management plan.

Two executive orders issued by former President Carter have become part of the federal coordination process intended to prevent environmental damage from federal actions. Executive Order 11988, issued on May 24, 1977, and entitled "Floodplain Management," directs each federal agency to "reduce the risk of flood loss, . . . minimize the impact of floods on human safety, health and welfare, and . . . restore and preserve the natural and beneficial values served by floodplains." Agencies are to incorporate floodplain planning into their decisionmaking and ensure that federal structures and facilities are consistent with the federal flood insurance and floodplain management programs described in Part III.

Executive Order 11990, issued on the same date as Executive Order 11988, is entitled "Protection of Wetlands." Each federal agency must "minimize the destruction, loss or degradation of wetlands, and preserve and enhance the natural and beneficial values of wetlands." In addition to doing wetlands planning, federal agencies must avoid undertaking or licensing construction in wetlands unless "there is no practicable alternative," and then all "practicable measures to minimize harm" must be taken.

There is no absolute prohibition on water resources projects in national parks, monuments, and wildlife refuges. However, there are specific statutes which deal with the protection of particular areas. In wilderness areas established under the National Wilderness Act, the president may approve a water resources project if he determines that it "will better serve the interests of the United States . . . than will its denial."

In addition to the section 404 program (to be discussed in Part III), several provisions of federal water pollution control law might place limitations on federal water resources development, assistance to private developers, and licensing. The Safe Drinking Water Act prohibits federal financial assistance for any project that may, through a recharge zone, contaminate an aquifer designated by the Environmental Protection Agency as a sole or principal source of drinking water for an area. Examples of proscribed federal financial assistance are grants, contracts, loans, and loan guarantees.

Section 401 of the Clean Water Act requires an applicant for a federal license or permit for any activity that may result in water pollution to obtain a certification from the state where the discharge originates that the discharge will comply with applicable effluent limitations and water quality standards. In contrast to the Coastal Zone Management Act, where a "consistency" determination is up to the federal agency, a state's denial of 401 certification or its conditional certification is binding on federal officials. This effectively gives states a veto over federal permits such as Corps 404 permits. Moreover, even though a federal project may be exempt from the requirement of obtaining a 404 permit, a state can control the discharge of dredged or fill material by federal agencies into navigable waters within the state unless the Corps claims interference with its navigation power, i.e., navigation cannot be maintained if the state conditions are met.

Under section 102(b) of the Clean Water Act, any federal agency planning the construction or financing of a reservoir shall consider the inclusion of storage for regulation of streamflow, but "storage and water releases shall not be provided as a substitute for adequate treatment or other methods of controlling waste at the source." The EPA administrator's comments on "the need for, the value of, and the impact of, storage for water quality control" must be a part of the project report submitted to Congress. The economic value of storage for regulation of streamflow is taken into account as a project benefit, but this value must not constitute a disproportionate share of project benefits. Costs of streamflow regulation features are reimbursable or nonreimbursable depending on whether beneficiaries can be identified. Where hydroelectric projects are concerned, there is a quantitative limit to the inclusion of storage for regulation of streamflow for water quality control.

Finally, where federal-aid highways are concerned, section 4(f) of the Department of Transportation Act prohibits the secretary of transportation from approving a project that requires the use of any publicly owned land from a public park, recreation area, or wildlife or waterfowl refuge or any land from an historic site unless (1) there is no feasible and prudent alternative to the use of such land, and (2) all possible planning has been done to minimize environmental harm. The term "use" means not only direct interference with public land but also indirect impacts such as noise or loss of access.

REFERENCES FOR PART II

1. Trelease, F. J. "Federal-State Relations in Water Law," National Water Commission, NTIS PB 203 600 (1971), p. 39.
2. Morreale, E. H. "Federal-State Rights and Relations," in *Waters and Water Rights*, R. E. Clark, ed. (7 vols., Indianapolis, IN: Allen Smith Co., 1967–1978), II, p. 9. The seven-volume treatise is hereafter referred to as "Clark."
3. Harnsberger, R. S. "Eminent Domain and Water," in Clark, IV, p. 114.
4. The Rivers and Harbors Act of 1970, P.L. 91–611, § 111, 84 Stat. 1818 (1970).
5. Hillhouse, W. A. "The Federal Law of Water Resources Development," in *Federal Environmental Law*, E. L. Dolgin and T. G. P. Guilbert, eds. (St. Paul, MN: West Publishing Co., 1974), p. 848.
6. Sax, J. L. "Federal Reclamation Law," in Clark, II, p. 121.
7. Worcester, D. *Rivers of Empire* (New York, NY: Pantheon Books, 1985), especially pp. 285–308.
8. *California v. United States*, 438 U.S. 645, 98 S. Ct. 2985, 57 L. Ed. 2d 1018 (1978).
9. Roos-Collins, R. "Voluntary Conveyance of the Right to Receive a Water Supply from the United States Bureau of Reclamation," *Ecology Law Quarterly* 13:773–878 (1987), p. 783.
10. Castleberry, J. N. "Federal Flood Control Activities," in Clark, IV, p. 222.
11. *Who Runs the Rivers? Dams and Decisions in the New West* (Stanford Environmental Law Society, 1983), pp. 134–166.
12. Castleberry, J. N. "Watershed Protection," in Clark, IV, p. 243.
13. Goldfarb, W., and J. Heenehan. "Legal Control of Soil Erosion and Sedimentation in New Jersey," *Rutgers Camden Law Journal* 11(3):379–422 (1980), p. 402.
14. Holmes, B. H. "History of Federal Water Resources Programs and Policies, 1961–1970," U.S. Department of Agriculture, U.S. Government Printing Office (1979), p. 71.
15. DeWeerdt, J. L., and P. M. Glick, eds. "A Summary-Digest of the Federal Water Laws and Programs," National Water Commission, U.S. Government Printing Office (1973), p. 73.
16. Butcher, W. R., P. R. Wandschneider, and N. K. Whittlesey, "Competition Between Irrigation and Hydropower in the Pacific Northwest," in Kenneth D. Frederick, ed. *Scarce Water and Institutional Change* (Washington, D.C.: Resources For The Future, Inc., 1986), pp. 41–42.
17. Castleberry, J. N. "The Federal Power Commission and Water Power Licenses," in Clark, IV, p. 301.
18. 354 F. 2d 608 (2nd Cir., 1965), *cert. den.* 384 U.S. 941.
19. Blumm, M. C. "A Trilogy of Tribes v. FERC: Reforming the Federal Role in Hydropower Licensing," *Harvard Environmental Law Review* 10:1–59 (1986), pp. 11–12.

20. Whittaker, M. C. "The Federal Power Act and Hydropower Development: Redis-covering State Regulatory Powers and Responsibilities," *Harvard Environmental Law Review* 10:135–187 (1986), pp. 142–143.
21. *Tulalip Tribes* v. *FERC*, 732 F. 2d 1451 (9th Cir., 1984).
22. Power, G. "The Federal Role In Coastal Development," in *Federal Environmental Law*, see 5 above, p. 813.
23. 33 CFR pt. 320 et. seq.
24. "Water Policies for the Future," National Water Commission, U.S. Government Printing Office (1973), p. 128.
25. Jaffe, A. B. "Benefit-Cost Analysis and Multi-Objective Evaluation of Federal Water Projects," *Harvard Environmental Law Review* 4:58–85 (1980), p. 70.
26. Ingram, H. M. "The Political Economy of Regional Water Institutions," *American Journal of Agricultural Economics* 55:10–18 (1973), p. 10.
27. Burton, J. S. "History of the Federal-State Water Resources Research Institute Program," *Water Resources Bulletin* 22:637–647 (1986).
28. 40 CFR pt. 1500 et. seq.
29. 437 U.S. 153, 98 S. Ct. 2279, 57 L. Ed. 2d 117 (1978).
30. 741 F. 2d 257 (1984).

PART III

Nontransformational Uses: Uses That Do Not Change the Waterbody

Law is an institutionalized means of resolving conflicts in socially acceptable ways. Thus, as conflicts in a particular sphere of social interaction increase so does legal activity. The burgeoning demand for nontransformational uses of water resources in recent years has intensified disputes among nontransformational users, e.g., competition for Colorado River rafting permits, and between nontransformational and transformational users, e.g., trout fishermen against dam builders. Heightened legal activity in these areas can be expected to continue.

Three categories of nontransformational use are discussed in this part: (1) recreational, aesthetic, and ecological uses of rivers, lakes, and shorelands; (2) floodplains and wetlands protection; and (3) water resource enhancement without additional water treatment.

Chapter 14

THE PUBLIC TRUST DOCTRINE

The public trust doctrine is the "wild card" of water law. It is an evolving, flexible concept that has been applied to a wide variety of resource uses, including many water-related uses discussed in this Part. The essence of the public trust doctrine, as articulated by Professor Sax, is that "(w)hen a state holds a resource which is available for the free use of the general public, a court will look with considerable skepticism on any government conduct which is calculated *either* to reallocate that resource to more restricted uses *or* to subject public uses to the self-interest of private parties."[1] Government is a trustee of all public natural resources, owing a fiduciary obligation to the trust beneficiaries—the general public—to maintain public uses unless diminishing them would achieve a countervailing public benefit.

What are the public natural resources that represent the "corpus" of the public trust? As inherited from English common law and refined by early American decisions, the public trust attached to the beds and banks of tidal and inland navigable waterways up to the high-water mark. The most famous public trust case in American law is the decision of the United States Supreme Court in *Illinois Central Railroad Company* v. *Illinois*.[2] In 1869 the Illinois legislature conveyed to the railroad a large parcel of submerged land representing virtually the entire Lake Michigan waterfront of the City of Chicago. The state received in return only the railroad's promise to pay the state a percentage of its gross earnings from wharves, piers, and docks to be built over the land. Four years later the legislature realized its mistake and repealed the conveyance without paying compensation to the railroad.

The U.S. Supreme Court upheld the repeal, concluding that the original conveyance was inconsistent with the state's public trust duties. A public trustee may transfer trust property for the public good, but it cannot abdicate its fiduciary responsibility so suddenly, so completely, and so unaccountably.

Since *Illinois Central*, federal courts have extended the public trust to national parks and their waterways and to fish and wildlife wherever located. Many state courts have also applied the public trust doctrine to water-related resource uses. In Massachusetts, filling of a Great Pond for a state highway has been enjoined. In Wisconsin, a ruling invalidated legislation that had authorized a private developer to drain a lake. Another Wisconsin decision prevented a local government from using a fishing stream for incompatible power generation. Wisconsin and California have used the public trust doctrine to protect inland shores and shoreside wetlands. California and New Jersey have based beach access rights on the public trust doctrine. In North Dakota, the Supreme Court has required state government to perform some short- and long-term water supply planning before granting large diversion permits. New Mexico and other western states have declared a public trust in groundwater in order to develop permit systems for evaluating proposals to export groundwater out of these states. And Montana has relied on the public trust doctrine to protect recreational uses of surface waters.

Just as the list of resources subject to the public trust has been expanded, so has the category of protected public uses. The early cases defined the public trust in terms of navigation, commerce, and fishing. But it is now settled that recreation, preservation of the existing ecological system, and aesthetic satisfaction also fall within the ambit of the public trust. Ecological and scenic preservation were the focus of the most important public trust case of modern times, *National Audubon Society* v. *Superior Court of Alpine City*.[3] Because of its extraordinary potential importance, especially in the West, this decision will be analyzed in detail.

Mono Lake, the second largest lake in California, sits at the base of the Sierra Nevada escarpment near the eastern entrance of Yosemite National Park. The lake is saline with no outlet. It contains no fish but supports large populations of brine shrimp that feed vast numbers of nesting and migratory birds. Mono Lake is also scenically and geologically important.

Historically, most of Mono Lake's water has come from snowmelt carried by five freshwater tributary streams arising in the Sierra Nevada. In 1940, however, the predecessor to the California Water Resources Board granted the city of Los Angeles a permit to appropriate virtually the entire flow of four of the five tributary streams. Since that time, Los Angeles has gradually diverted this water into the Owens Valley Aqueduct for water supply in the city.

As a result of these diversions, the level of Mono Lake has dropped, the surface area has diminished by one-third, and the lake's salinity has drastically increased. In the court's words, ''The ultimate effect of continued diversions is a matter of intense dispute, but there seems little doubt that both the scenic beauty and ecological values of Mono Lake are imperiled.'' On the other hand, ''the need of Los Angeles for water is apparent, its reliance on rights granted by the board evident, (and) the cost of curtailing diversions substantial.''

How does the public trust doctrine affect this clash between the degradation of ''a scenic and ecological treasure of national significance'' and the genuine water needs of a large city fulfilled by a long-standing appropriation permit? Before reaching this question, the California Supreme Court had to deal with two subsidiary issues: (1) Does the public trust doctrine apply to nonnavigable tributaries of a navigable waterbody? and (2) Is the public trust doctrine subsumed in the state water rights system or does it function independently of that system?

On the first question, the court held that under California law the public trust applies to activities on nonnavigable tributaries that interfere with public interests in navigable waterbodies—the Mono Lake situation. On the second question, the court held that the public trust doctrine is neither independent of nor subsumed under the appropriative water rights system:

> We are unable to accept either position. In our opinion, both the public trust doctrine and the water rights system embody important precepts which make the law more responsive to the diverse needs and interests involved in the planning and allocation of water resources. To embrace one system of thought and reject the other would lead to an unbalanced structure, one which would either decry as a breach of trust appropriations essential to the economic development of this state, or deny any duty to protect or even consider the values promoted by the public trust.

Therefore, an ''accommodation'' of the two systems must be forged. The principles of this accommodation are:

- The state retains continuing supervisory control over its navigable waters and the lands beneath those waters; no vested right may be acquired to appropriate water in a manner harmful to the interests protected by the public trust.
- Nevertheless, the state has the power to grant appropriative permits that may unavoidably diminish instream uses if the state's population maintenance and economic health outweigh protection of public trust uses.
- The state has an affirmative duty to take the public trust into account in the planning and allocation of water resources, and to protect public trust uses whenever feasible.
- Once the state has approved an appropriation, the public trust imposes a duty of continuing supervision over the taking and use of appropriated water. The state is not confined by past allocation

decisions which may be incorrect in light of current knowledge or inconsistent with current needs.

- The state accordingly has the power to reconsider allocation decisions even though those decisions were made after due consideration of their effect on the public trust.

Applying these principles to the Mono Lake situation, the court sent the matter back to the lower court and the California Water Resources Board to reconsider the appropriation permits to Los Angeles in light of ecological damage to Mono Lake. A key fact in this case was the Board's limited powers in 1940 when the permits were granted. At that time the Board had neither the power nor the duty, as it now has, to consider public trust interests in making allocation decisions. Temporal priority of actual appropriation for a beneficial use was the only concern in those days. Thus, with regard to the Mono Lake diversions, no responsible body "has determined that the needs of Los Angeles outweigh the needs of the Mono Basin, that the benefit gained is worth the price. Neither has any responsible body determined whether some lesser taking would better balance the diverse interests."

In California, after the Mono Lake decision, an appropriative diversion permit can always be contested as violating the public trust, even where the instream flow uses were considered in granting the permit. Will other states also allow ostensibly "vested" diversion rights to be reconsidered at any time when citizens assert public trust interests? Idaho has already adopted the Mono Lake rule. The role of the public trust doctrine in maintaining instream flows will be examined in Chapter 17.

The public trust doctrine is a convenient instrument for achieving changes in water law not only because of its adaptability to modified circumstances and public needs but also because it avoids the taking issue. Persons holding private property interests in public trust resources—such as tidelands owners or holders of California diversion permits (which have the status of property interests)—are presumed to have obtained their property subject to the trust rights of the public. Thus, they are not entitled to compensation when they are deprived of uses that conflict with the public trust because they did not have rights to these uses in the first place.

Chapter 15
PUBLIC USE
OF WATERBODIES

PUBLIC USE: OWNERSHIP OF BEDS AND BANKS

To a great extent, public use of waterbodies for fishing, boating, and swimming, and public use of banks for hiking, camping, and similar activities depends on whether the beds and immediate banks are owned by a state or by private riparian owners. Where a state owns the beds and banks, a waterbody is said to be ''public''—capable of public in-place use up to the limit of state ownership—the high-water mark of rivers and the low-water mark of lakes. These public rights are qualified in two major ways: (1) public users must legally gain access to the waterbody, and (2) public users must comply with state and local regulations regarding fish and game, noise, littering, etc.

In almost all states, beds and banks of tidelands and other navigable waterbodies are owned by the state in trust for the public. Even in those few states where beds and banks of nontidal navigable rivers are owned by riparians, the result is the same because the riparian owner holds the beds and banks subject to public rights of navigation and in-place use.[4] The concept of ''navigability'' is one of the most meaningful and difficult elements of water law. As shown in Part II, the limits of federal jurisdiction over waters based on the Commerce Clause, i.e., federal regulation and the navigation servitude, are determined under a federal navigability test: Is the waterway tidal or suitable for interstate commerce? ''Navigability'' also governs ownership of beds and banks, and this second navigability test is also a matter of federal law, but there are significant differences between this "title test" and the "jurisdiction test" discussed in Part II. In the first place, the title test involves navigability at the time of the formation of the Union in the original states or the admission to statehood of those formed later, not whether the waterbody can be made navigable by artificial improvements.[5] Second, under the title test navigability in *intrastate* commerce is all that is required, not usability in interstate commerce. The first difference is more important where rivers are concerned: potential navigability after dredging or channelization will not satisfy the title test, although it will satisfy the jurisdiction test. The second difference is more important where lakes are concerned: large, isolated, intrastate lakes might satisfy the title test but not the jurisdiction test. Both tests liberally interpret "commerce" in terms of fur traders' canoes or commercial log floating. In summary, for the purpose of determining title, the question of navigability is a federal question, to be determined by the "federal test" of whether the waterbody in its natural and ordinary condition was usable for intrastate commerce by the customary modes of trade or travel on water when the state was admitted to the Union.

In Wisconsin and Iowa, the state retains title to the beds of some waterbodies that are navigable under a state law test even if they are not navigable under the federal test. But the overwhelming majority rule is that riparian owners on rivers that are nonnavigable under the federal title test own up to the "thread" (center line of the main channel) of a river. (Ownership of lakebeds will be discussed in Chapter 16.) The derivation of this rule is as follows:

> Historically, public rights attached to navigable waters because of the public need to use such waters. Since the basic public use was navigation, it was assumed that waters not susceptible to navigation were not required for public use. As a consequence, there were no public rights in nonnavigable waters, their beds, or banks.[6]

Thus, the classic legal position was that the public had no rights to use waterways nonnavigable under the federal title test.

The strict classic rule has been modified or abolished in every state. But each state has adopted its own strategy for allowing at least some public use of waterbodies that are nonnavigable under the federal title test.[7] As previously mentioned, Wisconsin and Iowa claim ownership of beds and banks of certain waterbodies that pass a state test of navigability. Among all the other states, where riparians own the beds and banks of waterbodies nonnavigable under the federal test: some states bar wading and fishing from a boat but allow floating on the theory that water cannot be privately owned, but is owned by the state in trust for the public; other states prohibit wading but allow fishing from a boat because fish are also held in trust by the state; and still other states bar wading but allow contact with the streambed for pushing or pulling a boat. In many states, particular nonnavigable rivers have become public because of long-term public use ("prescriptive easement").

The most interesting development toward loosening the classic restriction on public use of waterbodies nonnavigable under the federal title test is the tendency of an increasing number of states to redefine "navigability" under state law to support public uses. Courts in these states have decided that although riparians may own the beds and banks, public rights exist in waterbodies that are "navigable in fact," "floatable," or "suitable for pleasure craft." We have here a third definition of "navigability"—this time under state law—that emphasizes recreational navigation and other public uses. And where the waterway itself has become public, the public is also permitted to make reasonable use of beds and immediate banks. According to Professor Stone, the trend is toward opening for public use all waters suitable for public use.[4] States that have adopted the "recreational boatability" test of navigability, or some close variant of it, are: Arkansas, California, Georgia, Idaho, Massachusetts, Minnesota, Missouri, Montana, New York, Ohio, Oklahoma, Oregon, Wisconsin, and Wyoming.[8]

PUBLIC ACCESS

Although a waterbody may be usable by the public despite the fact that its bed and banks are privately owned, public usability is academic without public access. In most cases, access to public waterways through private property must be purchased, leased, or acquired by eminent domain from the landowner, and in these days of depleted government coffers, this may be tantamount to no public access at all.

Fortunately, there are many instances where access to public waterbodies may be procured without governmental outlay. A river may be reached from a public right of way for a road or railroad; once there, a user can walk or wade below the high-water mark. Canoeists may gain access at upstream public property and float down through private property, beaching their canoes and using the immediate banks for picnicking and camping. In addition, there are state legal doctrines mitigating the general rule that the public has no right to cross private land to reach public waters. New England Great Ponds are not only public waters, but the public is given rights to cross certain private lands in order to reach them. Access may also be simplified in states once covered by the Northwest Ordinance.

Prescriptive easements and implied dedication may also be relied upon to establish public access. An "easement" is a legal right to use another's land, most often for passage. Where the public generally has used a pathway continuously, openly, and adversely to the landowner for the prescriptive period, the public may acquire an easement to continue that use. Commonly, a landowner's defense to the claim of a public prescriptive easement is to assert that the public use was permissive, not adverse. But permissive use may give rise to an implied dedication of access to the public:

> Dedication is the donation of land, or of rights in land, to the public for the use of the public. It requires only an intent on the part of the landowner . . . to dedicate and an acceptance on the part of the public. Unlike prescription, no minimum period of public use is required, . . . And since an essential element is the intent that the public have the use of the land, dedication is necessarily permissive in character; the critical question is likely to be whether by his acquiescence the landowner intended a permanent dedication or only a temporary, revocable license.[4]

Actual public use is the key to proving an implied dedication, but it is also helpful if public funds have been used to maintain the disputed right of way, or if the landowner has fenced the area off from his other lands.

Under the Submerged Lands Act, coastal states own the beds of the marginal seas out to three miles or, in the cases of Texas and the Gulf coast of Florida, three marine leagues, which is approximately ten and one-half miles seaward from the ordinary low-water mark. The dividing line between the landward ownership of the states and the seaward ownership of private riparian owners is generally the mean high tide line, although Delaware, Massachusetts, Maine, New Hampshire, and Virginia use the low-water mark as a demarcation point.[5] Thus, in most coastal states the state itself owns the "foreshore," the wet sand area between the high and low water marks. The majority rule is consistent with the doctrine that states own the beds and banks of navigable waterways, with any area subject to tidal action presumed to be navigable.

A state's tidelands ownership is subject to the public trust. Public access to beaches has been achieved, in various states, through the public trust doctrine, prescriptive easement, express and implied dedication, local custom, constitutional provision, and state statute.

Chapter 16
USE OF LAKES

On lakes that are navigable under the federal title test, a riparian owner's property extends to the low-water mark. Where the level of a navigable lake fluctuates, the public has a right to go where the navigable waters go, even where they overlie privately owned land. Thus, a lakeshore owner on a navigable lake cannot fill his shoreland below the high water mark so as to impede the public's right of navigation.

Lakebeds of nonnavigable lakes belong to the riparian owners, except in those few states that have retained title to the beds of lakes nonnavigable under the federal title test. Determining the boundaries of lakeshore property on a nonnavigable lake is far more difficult than ascertaining the riverbed rights of a riverine riparian:

> Hopefully the lake may be round, in which case it can be divided as a pie; or it may be elongated with a fairly regular shoreline, in which case it can be divided down the middle, with just the ends divided as a pie. But if it is odd-shaped, with bays, inlets, promontories, and arms, and perhaps a sprinkling of privately-owned islands, the trial judge will not only have to exercise the wisdom of Solomon, but also have to execute his decree by dividing the baby.[4]

Allocating ownership of a lakebed among lakeowners may be important for a number of reasons. A lakeowner may wish to extract minerals from his portion of the lakebed; or if the lake is drained it will be necessary to plot boundary lines. Moreover, although it may sound absurd, in some states lakeowners have exclusive recreational rights to the lake surface overlying their lakebeds. In these states, one lakeowner can prevent others from using his piece of the pie. The more modern and increasingly popular rule is that lakeowners share in common the use of the lake surface as mutual ease-

122

ments over one another's lakebeds. One lakeowner may use the entire lake surface as long as his use is reasonable (i.e., does not interfere with the reasonable uses of other lakeowners).

ARTIFICIAL LAKES

Artificial lakes can be created in three ways:[9] (1) where a wholly artificial waterbody is established, unconnected with any previously existing natural body of water, such as where a quarry fills with rainwater or a developer pumps groundwater to fill an excavation; (2) where an existing waterbody is expanded, covering land previously dry, generally by damming a stream; and (3) where two existing lakes are connected by an artificial channel.

Wholly artificial lakes have caused few legal problems because of their relatively small size and typical location within a single landholding. But expansion of existing waterbodies often has significant impacts on other landowners and sometimes on the general public.

The prevailing rule is that there are no riparian rights in artificial lakes.[8] In accordance with this rule, the Florida Supreme Court has held that an owner of land that abuts or underlies part of a man-made, nonnavigable lake, created by damming a stream, is not entitled to the recreational use of the surface of the entire lake.[10] The Michigan Supreme Court has extended the rule to artificial extensions of existing natural lakes. In *Thompson* v. *Enz*,[11] a land development corporation purchased a parcel of land which had 1,415 feet of frontage on a 2,680-acre navigable lake that was used primarily for recreational purposes. The corporation subdivided a parcel into approximately 150 lots, only 16 of which abutted directly on the lake. The developer was to dig a canal to provide lake access for back lot owners. Before the canal had been completed, other lakeshore owners sued to enjoin the development for alleged violation of their riparian rights. The Court held that (1) since riparian rights do not attach to an artificial watercourse, and the canal would be an artificial watercourse, back lot owners did not possess riparian access rights; and (2) since riparian rights are not transferrable apart from riparian land, the developer could not have legally transferred access rights to the back lot owners. The case was "remanded" (referred back) to the trial court for consideration of whether the developer could have given the back lot owners permission to use its access rights, but the Michigan Supreme Court suggested (and the trial court later held) that permission was unreasonable in light of the lake's limited carrying capacity. A number of states, among them Idaho, Michigan, Washington, and Wisconsin, have enacted statutes requiring permits for lakeshore development that might injure ecosystems or other lakeshore owners.

Where a nonnavigable stream has been dammed to create a lake that is navigable-in-fact, public rights to the surface do not arise.[9] The general public should not possess rights that other lakeshore owners do not.

In a number of cases, lakeshore owners have sued to enjoin the draining of lakes formed by damming streams. Is there a right to the maintenance of an artificial water level? Some courts refuse to recognize such a right because the lakeshore owners have no riparian rights in artificial lakes. Other courts have granted injunctions based on prescriptive easement or dedication theories. A Florida statute declares it unlawful to drain any lake over two square miles in areas outside the Everglades without the written consent of all lakeshore owners. A number of other states prohibit lake drainage without a permit.

So diffuse is decisional water law on the issue of lake-level maintenance that Florida and a number of midwestern states have enacted lake-level maintenance statutes.

> They are designed to perform two basic functions. First, establishment of a level through a statutory procedure quantifies the . . . owner's common law right. Second and related to this, the statutes provide some protection for . . . owners for property damage caused by dams and other structures which result in a substantial alteration of the level of a lake.[5]

These statutes vary widely as to form of proceeding (administrative or judicial), waters covered (public or private), and standards for fixing lake levels ("average normal water level," "natural ordinary high water level," or "normal height and level"). Once lake levels are set, these statutes provide sources to fund the reconstruction or maintenance of dams.

With regard to an artificial connection between two lakes, most of the litigation has centered around public access to the artificial channel or canal.[9] A wholly artificial canal built on private property is not available for public use, even though the canal is navigable-in-fact or connected to a navigable waterway. Where the artificial canal results from the expansion of a natural channel, the public will have access if the channel was originally navigable, but will not have access if the channel was originally nonnavigable.

In the case of *Kaiser Aetna* v. *United States*,[12] the United States Supreme Court confronted the issue of whether the public should have access to a lake connected to the ocean by an artificial channel. Kuapa Pond on Oahu, Hawaii, was a shallow lagoon contiguous to Maunalua Bay but separated from it by a barrier beach. Ponds such as Kuapa have traditionally been considered private property under Hawaiian law. In 1961 a company called Kaiser Aetna bought the pond and surrounding lands for a subdivision development to be known as "Hawaii Kai." Kaiser Aetna dredged and filled parts of the pond, erected retaining walls, and built bridges within the development to create the Hawaii Kai Marina. Subsequently, Kaiser Aetna dredged an 8-foot-deep channel connecting Kuapa Pond to Maunalua Bay and the Pacific Ocean. The Army Corps of Engineers had been informed of Kaiser Aetna's activities, but had responded to Kaiser Aetna that no permits were necessary under section 10 of the Rivers and Harbors Act.

As a result of Kaiser Aetna's development activities, a marina-style community of approximately 22,000 persons surrounds Kuapa Pond. It includes approximately 1,500 waterfront lot lessees. The waterfront lot lessees, along with a number of nonmarina lot lessees from Hawaii Kai and some nonresident boat owners, pay fees for maintenance and security of the pond. Kaiser Aetna restricts access to the pond from the bay to boats belonging to members and their guests.

In 1972 the Corps demanded that Kaiser Aetna provide public access to the pond, claiming that as a result of improvements it had become a navigable water of the United States. When Kaiser Aetna refused to grant public access, the United States sued for an injunction in federal court. Two lower federal courts disagreed on the public access question.

The U.S. Supreme Court rejected the government's contentions. Justice Rehnquist, writing for a six-judge majority, began by differentiating between "navigability" and the "navigational servitude" with its "no compensation rule." In the Court's view, the federal navigational servitude is not applicable to all navigable waterways for all purposes. As for Kuapa Pond in particular,

> It is clear that prior to its improvement Kuapa Pond was incapable of being used as a continuous highway for the purpose of navigation in interstate commerce. Its maximum depth at high tide was a mere two feet, it was separated from the adjacent bay and ocean by a natural barrier beach, and its principal commercial value was limited to fishing. It is not the sort of "great navigable stream" that has previously been recognized as being "incapable of private ownership" . . . And, as previously noted, Kuapa Pond has always been considered to be private property under Hawaiian law. Thus, the interest of petitioners in the now dredged marina is strikingly similar to that of owners of fast land adjacent to navigable waters.

Even though Kuapa Pond is now navigable, it is not subject to the federal navigational servitude because it is an artificial lake that was not originally navigable.

The Court then concluded that the government's elimination of Hawaii Kai's "right to exclude" the public would amount to an unlawful taking of property for which just compensation would have to be paid according to the Fifth Amendment to the U.S. Constitution. Public access would cause an "actual physical invasion of the privately owned marina" tantamount to a taking:

> Thus, if the Government wishes to make what was formerly Kuapa Pond into a public aquatic park after petitioners have proceeded as far as they have here, it may not, without invoking its eminent domain power and paying just compensation, require them to allow free access to the dredged pond while petitioners' agreement with their customers calls for an annual $72 regular fee.

The majority was obviously impressed by Hawaii Kai's large investment and

the Corps' past condonation of Kaiser Aetna's water resources projects at
Hawaii Kai.

In a dissenting opinion, Justice Blackmun wrote that it does not
"advance analysis to suggest that we might decide to call certain waters
'navigable' for some purposes, but 'non-navigable' for purposes of the navigational servitude. . . . In my view, the power we describe by the term 'navigational servitude' extends to the limits of interstate commerce by water;
accordingly I would hold that it is coextensive with the 'navigable waters
of the United States'." For the three dissenters, "the developers of Kuapa
Pond have acted at their own risk and are not entitled to compensation for
the public access the Government now asserts."

Chapter 17
USE OF RIVERS AND STREAMS

WILD AND SCENIC RIVERS PROTECTION

In 1968 Congress passed the Wild and Scenic Rivers Act declaring it to be "the policy of the United States that certain selected rivers of the Nation which, with their immediate environments, possess outstandingly remarkable scenic, recreational, geologic, fish and wildlife, historic, cultural, or other similar values, shall be preserved in free-flowing condition, and that they and their immediate environments shall be protected for the benefit and enjoyment of present and future generations." This act established the wild and scenic rivers system, which now includes 72 rivers or river segments totaling 7,363 miles.

Rivers are admitted to the system in two ways. The legislature of a state through which a river flows may designate the river as wild, scenic, or recreational and request the governor to petition the secretary of the interior for inclusion in the federal system. The secretary can only consent if the river is to be permanently administered by the state "without expense to the United States other than for administration and management of federally owned lands." Land and Water Conservation Act funds are not considered as an "expense to the United States." Because states have so little to gain by this method of inclusion in the federal system, and so much to lose by foreclosing options with regard to a river, only 12 rivers totaling 753 miles have been added to the system by state designation. Proponents of river preservation recommend that Congress establish a separate grants program to stimulate state designation of rivers.

Most rivers in the federal wild and scenic rivers system have been included by act of Congress. A number of "instant" river components were designated when the Wild and Scenic Rivers Act was passed in 1968 and when the Alaska National Interest Lands and Conservation Act was enacted in 1980. Nowadays, Congress must first designate a particular river or segment as a "study river." (Occasionally, Congress votes a river into the system without a formal study having been made, as in the case of the New River in North Carolina.) Then, the secretary of the interior, through the National Park Service (NPS) which administers the system, performs a study and reports to the president on the suitability of the river for inclusion in the system. Before submitting its report to the president, the NPS must solicit comments from the federal water resource development agencies and the governor of each state involved. On the basis of the report and comments, the president may recommend designation to Congress, which may then pass another act admitting the river to the system. This process is so protracted and cumbersome that, for the most part, comparatively few noncontroversial rivers are designated. Ironically, congressional nomination as a study river may result in destruction of the very values that made the river worthy of potential admission to the system. Since there are no restrictions on private activities during the study period, lasting 10 years in some cases, land speculation, development, and tourism—enhanced by increased publicity—exacerbate existing threats to a river's wild, scenic, and recreational resources.

In passing a river into the federal system, Congress names a federal management agency that is required to determine the boundaries of the river component, decide whether it is to be managed as a wild, scenic, or recreational river, and prepare a management plan. However, the boundaries of the river component are limited to 320 acres per mile, approximately one-fourth mile, on both sides of the river. The Wild and Scenic Rivers Act makes no attempt to protect a river from the adverse effects of private development in the watershed beyond the corridor.

Even within the corridor, federal management authority is limited. The act contains no authority for federal agencies to regulate, by zoning or otherwise, private land uses within the corridor. Moreover, there are strict acreage, percentage, and other restrictions on federal condemnation of fee simple interests in the corridor. (A "fee simple interest" is the entire bundle of rights that we call land ownership.) Congress placed these restrictions on fee condemnation in order to placate local governments, which traditionally complain of a federal "land grab" when they lose property tax ratables by condemnation. Instead, Congress stressed the purchase or condemnation of "scenic easements"—rights to develop river corridors—which are less than fee simple interests and accordingly leave "title" in the landowner. But scenic easements are almost as expensive as fee interests, and funds for the acquisition of easements are scarce. Because the threat of federal condemnation is frequently hollow, a wild or scenic river is inadequately protected against private development that jeopardizes its integrity. Local governments are generally more interested in development than preservation.

In contrast, a federal wild, scenic, or recreational river is protected against incompatible federal activities by section 7(a) of the act:

> The Federal Power Commission [now the Federal Energy Regulatory Commission in the Department of Energy] shall not license the construction of any dam, conduit, reservoir, powerhouse, transmission line, or other project or works . . . , on or directly affecting any river which is designated . . . as a component of the national wild and scenic rivers system or which is hereafter designated for inclusion in that system, and no department or agency of the United States shall assist by loan, grant, license, or otherwise in the construction of any water resources project that would have a direct and adverse effect on the values for which such river was established, as determined by the Secretary charged with its administration. Nothing contained in the foregoing sentence, however, shall preclude licensing of, or assistance to, developments below or above a wild, scenic or recreational river area or on any stream tributary thereto which will not invade the area or unreasonably diminish the scenic, recreational, and fish and wildlife values present in the area.

Study rivers are afforded the same protection during the first six years of the study period. Federal lands within or adjacent to the corridor must be managed in accordance with the management plan developed by the managing federal agency.

Because of its lack of authority for federal control of private activities inside and outside river corridors, the federal Wild and Scenic Rivers Act has been more successful in preserving rivers that flow through federal land. An alternative to inclusion of a river in the federal wild and scenic rivers system is utilization of a state rivers preservation program. Twenty-eight states have established statewide river conservation programs, protecting 327 river segments totaling almost 12,000 miles. All state programs provide some degree of state-level protection against the construction of dams. Most state programs also provide protection against massive, dewatering diversions and against channelizations. However, "most states are only marginally successful in controlling development within designated river corridors."[13] California, Maine, Michigan, New York, Oregon, and Washington have some sort of statewide zoning, and all but Washington use it for river corridor management. Furthermore Maine, Minnesota, Oregon, Vermont, and Washington administer statewide shoreland zoning programs. The majority of states, however, use inconsistent and unreliable local zoning to protect river corridors.

No state rivers protection statute can conclusively protect a river against federal development activities or private development under federal license. But, as was pointed out in Chapter 11, small hydropower incentives for new dams are unavailable on rivers or stretches that are included in, or are being studied for inclusion in, state river protection systems. Another major advantage of the federal process is that designation of a wild, scenic, or recreational river creates an implied federal reserved right for instream flow maintenance with priority as of the designation date. This is true even though states otherwise control water diversions on federal wild and scenic rivers.

RIVER CORRIDOR PROTECTION

Wild and scenic rivers statutes and state shoreland zoning programs are but two of many legal strategies that may be utilized to protect sensitive river corridors. At the federal level, the federal development and licensing programs, as well as the environmental limitations on these activities, all of which have been discussed in Part II, might be used as parts of a comprehensive strategy to manage river corridors. The National Flood Insurance Program (see Chapter 19) and the federal wetlands protection programs (see Chapter 20) might also be incorporated into this strategy. However, since the demise of the Water Resources Council and its river basin commissions (see Chapter 12) there has been little effective coordination among federal agencies with diverse legal and political agendas.[14]

In addition to wild and scenic rivers and shoreland zoning programs, states can use the following other programs to manage river corridors: floodplain management; wetlands protection; critical and special management areas (e.g., Minnesota and Wisconsin have passed special statutes for the St. Croix River; Maine has adopted a special corridor management statute for the Saco River; and Florida has adopted a special critical area designation for the Apalachicola River); dam regulation; endangered species protection; acquisition of fee simple titles and conservation easements; water quality control; water diversion management; watershed management; regulation of natural resources extraction; fish and wildlife management; regulation of hazardous and solid waste disposal; parks and natural areas management; facilities siting; and agricultural lands preservation.[13,14]

The state of Oregon has adopted unique measures for protecting its riverine areas. First, an owner of riparian land who agrees to protect or restore streamside vegetation is eligible for a complete property tax exemption for qualifying lands. Second, Oregon law provides for a state personal income tax credit for up to 25% of the costs incurred in a fish habitat improvement project certified by the state.

According to Kusler,[14] "some of the most innovative corridor management efforts are at the local level." Greenways, stream corridor protection ordinances, and river corridor commissions are some of the devices that local governments use to protect river corridors.

INSTREAM FLOW PROTECTION

Assurances of public usability and public access are futile if the streamflow is inadequate for nontransformational use. Consequently, legal protection of minimum instream flows has expanded commensurate with increased recreational and environmental demands on water resources.

In those eastern states where the riparian doctrine of water diversion is still in effect, there is some degree of protection for instream uses "as an incidental consequence of its general allocation principles."[15] A riparian owner's rights to water-based recreation and aesthetic satisfaction have equal status with the transformational uses of other riparians. This balance tends to preclude dewatering of streams. Moreover, the riparian doctrine's limitations on diversions off riparian land and outside the watershed serve to protect instream uses. But

> (u)nder the riparian doctrine, water allocation decisions consist of court decisions for resolving water-use disputes, an approach providing discontinuous and somewhat random attention to the water-use situation. In the absence of a riparian landowner with interest in instream activities, significant harm may occur to instream values without attention from the courts. Even where lawsuits are initiated, attention may be focused on competing offstream water needs with little attention given to instream uses. Opportunity for state agency involvement may be limited. Similarly, members of the public with interest in instream values may not have proper standing to initiate court proceedings or to participate in cases involving riparian landowner conflicts. Thus, consideration of instream water uses under the riparian doctrine may depend on the fortuitous location of a riparian landowner interested in maintaining such uses.[15]

Moreover, fish and wildlife do not possess riparian rights.

Eastern permit states have tended to grant administrative agencies authority to set minimum streamflows. Criteria for establishing minimum flows in these states range from vague "public interest" standards to more descriptive standards (e.g., "historical or reasonable flows below dams," "average flow," "average minimum flow," "normal flow," "optimum flow") to formulae (e.g., "the average minimum flow for seven consecutive days within the lowest flow year of record," "twenty-five percent of natural streamflow").

Federal and state regulatory water rights to instream flow maintenance are also important in the East. Some of the relevant federal statutes here are the Federal Power Act (the FERC can set minimum flows in hydropower licensing proceedings), the Endangered Species Act, the Wild and Scenic Rivers Act, and Sections 401 ("state certification") and 404 ("dredge-and-fill") of the Clean Water Act. The National Environmental Policy Act and the Fish and Wildlife Coordination Act require that streamflows be considered when the federal government develops or licenses a water resources project. Finally, federal reserved rights to minimum flows for national parks, national forests, national wildlife refuges, and wild and scenic rivers may some day become as important in the East as they have been in the West.

Because much of the West is arid, instream flow issues have been more compelling there than in the East. Professor Tarlock has identified direct and indirect methods of flow preservation in the West.[16] Direct measures of dedication are (1) flow reservation systems; (2) flow appropriation systems; (3) administrative review of new diversion permits, and denial or conditioning of permits based on the need for instream flows; and (4) federal reserved rights for instream uses. Indirect methods include regulatory water rights, limited term permits, vigorous prosecution of forfeiture and abandonment cases, and conservation.

Six states (Alaska, Iowa, Kansas, Montana, Oregon, and Washington) have adopted streamflow reservation systems, where a basic level of streamflow is set aside below which no new water rights can be granted.[17] Montana has used this approach to reserve future water flows in the Yellowstone Basin for instream flows (5.5 million acre-feet), irrigation (655,000 acre-feet), municipal use (61,000 acre-feet), and offstream storage (1.2 million acre-feet). Colorado, Idaho, and South Dakota have developed flow appropriation systems where the state is granted a water right for a given instream use. The advantage of this approach is that the state, as a junior appropriator, can object to changes in use (e.g., upstream transfers of diversion points) that will detrimentally change stream conditions on the portion of the stream for which the minimum flow appropriation is held. A major disadvantage of this approach is that some states still require physical diversions to perfect appropriations. Colorado has enacted a statute obviating physical diversions for instream flow appropriations, and Idaho has accomplished this result through court decisions.

Among the western states, Alaska, California, and Montana authorize administrative agencies to condition new water rights on protecting values that are in the public interest, including flow maintenance. In 1985, the Montana legislature enacted a statute infusing a range of public interest considerations into the state water code, with criteria being especially rigorous for diversions of over 4,000 acre-feet. The act prohibits private appropriations for transport out of specified basins or for any consumptive use in excess of 4,000 acre-feet. Only the Department of Natural Resources and Conservation may make these kinds of appropriations. The only way in which a private user can transport water out of the specified basins or obtain rights to consume more than 4,000 acre-feet is to lease water from the department.

Federal reserved rights to instream flows have significant impacts on flow maintenance in the West. Indian reservations are generally located on downstream stretches. In addition, federal reserved rights for Indian agriculture and fishing carry comparatively early priority dates. National parks, forests, and wilderness areas, on the other hand, are customarily located in headwaters areas where there is little competition with private appropriators. Moreover, these federal areas carry reserved rights with comparatively recent priority dates. Where national forests are concerned, federal reserved rights to instream flows have been limited by the United States Supreme Court's decision in *U.S.* v. *New Mexico*.[18] The Court there held that the Forest Service cannot claim reserved rights for instream flows where forest and watershed protection are not directly involved; fish and wildlife preservation and aesthetics are not legitimate grounds for asserting reserved rights to instream flows for national forests.

With the exception of federal reserved rights, the direct approaches outlined above are futile where a waterbody is already fully appropriated, as is the case in many areas of the West. Under pure appropriation theory, an appropriator can completely dewater a stream if the water can be used beneficially. The number of dry, or nearly dry, streambeds in the West during late summer and early fall attests to the success of the appropriation system in facilitating diversion of all available water for economically gainful pursuits. This common situation militates against the use of flow reservation, flow appropriation, or new permit review to preserve instream flows. In many parts of the West, their adoption would be "too little, too late." Senior appropriators' rights could not be disturbed without paying compensation.

One means of escaping this dilemma might be the assertion of federal regulatory water rights, and the negation of the "taking issue" through the navigation servitude. But the U.S. Supreme Court has warned that the federal navigation power may not always be coextensive with the navigation servitude (see Chapter 16).

Another promising approach to maintaining instream flows is the public trust doctrine. As discussed in Chapter 14, property owners of land included in the public trust are presumed to hold their titles subject to the trust; thus, if the public trust is asserted, they have not really "lost" anything. After the Mono Lake decision, states may invoke the public trust doctrine to reclaim instream flows on overappropriated waters without paying compensation to existing rightsholders. Professor Ausness feels that this approach would be unnecessary and unfair.[19] However, it might also be argued that not only is the public trust approach necessary to restore and maintain instream flows on many western streams, but that it is the general public—deprived for so long of its recreational and environmental rights—which is deserving of compensation.

COMPETING NONTRANSFORMATIONAL USES

Conflicts between nontransformational and transformational river users have been common for many years; e.g., disputes over instream flow protection, attempts to forestall or modify water resources development using one of the federal statutes mentioned in Part II or a state counterpart, and water pollution control efforts. But a relatively recent phenomenon is the competition among nontransformational users for river recreation. These conflicts arise both between and within use classes. For example, boat and bank fishermen may lobby for restrictions on canoeing in a particular river because the fishermen claim that canoeists are disturbing them and the fish. Within the fishing fraternity, fly-fishermen frequently seek to exclude bait and spin fishermen from a river by recommending the imposition of "fly-fishing only" or "catch-and-release" regulations. These internecine battles among nontransformational users can only intensify as recreational pressure on river resources increases.

In this context, the case of *Wilderness Public Rights Fund* v. *Kleppe*[20] was a harbinger of the future. There, the U.S. Court of Appeals for the Ninth Circuit was called upon to determine the validity of a National Park Service decision allocating user days on the Colorado River between commercial and oncommercial rafters. Access to Colorado River rafting has been limited since 1972. No raft trip may be made without a NPS permit, and river use had been frozen at the 1972 level—92% allotted to commercial concessioners who lead guided trips for a fee, and 8% to noncommercial users who apply for permits as private groups. The plaintiff, Wilderness Public Rights Fund (WPRF), argued that its members were being unfairly treated, and advocated a lottery or first-come-first-served permit system which would allow successful permittees to choose between joining a guided party or a noncommercial party. Is it fair, asked the WPRF, that skilled noncommercial applicants be compelled to plan their trips years in advance while neophytes who make guided trips can contact concessioners and make arrangements at the last minute? Before the court could render its decision, the NPS changed its allocation to 70% of user days for commercial trips and 30% for noncommercial trips.

Chapter 18
FEDERAL OUTDOOR RECREATION PROGRAMS

Outdoor recreation has experienced phenomenal growth since the end of World War II, and roughly half of all outdoor recreation is water oriented.[21] The federal outdoor recreation program is the focal point of a national network of federal, state, county, and municipal public recreation sites, many featuring water-related outdoor activities.

The secretary of the interior is authorized to (1) inventory and evaluate the nation's outdoor recreation needs and resources, (2) prepare a nationwide outdoor recreation plan identifying problems and recommending solutions, (3) provide technical assistance to public and private interests, and (4) promote coordination of federal plans and activities relating to outdoor recreation. The National Park Service has replaced the Heritage Conservation and Recreation Service and its predecessor, the Bureau of Outdoor Recreation, as the organ within the Department of the Interior that is responsible for federal outdoor recreation planning.

Budget cuts have sharply curtailed not only federal outdoor recreation planning but also implementation of plans through the Land and Water Conservation Fund. The Fund is made up of congressional appropriations, admission and user fees from federal outdoor recreation areas, and certain other receipts. In general, 60% of the fund is available for grants to states on a 50–50 matching basis for planning, acquisition, and development of land and water areas, and 40% for acquisitions by the National Park Service, Forest Service, and Fish and Wildlife Service. No state may receive funds unless it has a comprehensive statewide outdoor recreation plan (SCORP)

135

approved by the Department of the Interior. A sizable portion of a state's share is passed through to local governments. "About three-fourths of the Federal funds and two-thirds of the State and local funds have gone to water-oriented recreation."[21]

Nearly 700 reservoirs constructed by the Army Corps of Engineers and the Bureau of Reclamation have public recreation facilities. These facilities are generally administered by the construction agency itself, but occasionally by the National Park Service, the Forest Service, or the Fish and Wildlife Service. The National Water Commission has pointed out that construction agencies do a notoriously poor job of administering recreation. In light of this failure, Congress enacted the Federal Water Project Recreation Act of 1965, stipulating that (1) nonfederal bodies must agree to administer recreation and pay one-half of the construction costs allocated to recreation and all the maintenance, operation, and replacement costs for recreation; (2) without an agreement the construction agency can build only minimum recreation facilities for health and safety, plus acquire potential recreation land and hold it for 10 years pending negotiation of the required cost-sharing agreement; and (3) at the end of 10 years, if no agreement has been reached, the acquired land must be sold. However, this statute has been more honored in the breach than the observance.[21] The Soil Conservation Service was also faulted by the National Water Commission for failing to arrange for public access to thousands of small watershed projects that it finances.

Since 1916 the National Park Service has managed public land for preservation and recreation, including water-based recreation. The NPS administers the national parks, national recreation areas, national monuments, national seashores, national lakeshores, national rivers, national wild and scenic rivers, the Boundary Waters Canoe Area in Minnesota, national gateway parks, and the Big Thicket and Big Cypress national preserves. The major legal problem confronting the NPS is whether and to what extent it can regulate activities on inholdings and adjacent lands that are inimical to the purposes of the parks without having to pay compensation. For example, can the NPS regulate logging activities on private lands outside Redwood National Park where the private logging causes siltation of the streams in the park?

Under the Wilderness Act, Congress has set aside some 19 million acres for preservation in the National Wilderness Preservation System. Wilderness areas are managed either by the NPS or the Bureau of Land Management in the Department of the Interior, or the Forest Service in the Department of Agriculture. Water resource projects cannot be built in wilderness areas without presidential approval. No new mining claims may be filed in wilderness areas, but existing claims may be worked subject to environmental restrictions. Hunting and fishing are permitted in wilderness areas, but no motorized equipment may be used.

The Forest Service administers the 190 million acres of the National

Forest System, and the Bureau of Land Management the 450 million acres of public lands, under a statutory standard of "multiple use, sustained yield." The multiple uses that must be delicately balanced by the management agency are outdoor recreation, range, timber, watershed, fish and wildlife, and scenic-archeologic. Some land will be used for fewer than all of these resources, but the potential for using each relevant resource must be preserved. The multiple use criterion is so ambiguous that it places virtually no legal constraints on a management agency. For this reason, Congress enacted the National Forest Management Act, requiring the Forest Service to protect water resources from the adverse effects of clearcutting.

Finally, the Fish and Wildlife Service in the Department of the Interior oversees over 30 million acres of the National Wildlife Refuge System. Although managed primarily for fish and wildlife conservation, hunting, fishing, and recreational boating are permitted as long as these activities are consistent with fish and wildlife management. In one case, a federal district court prohibited powerboat use in a wildlife refuge. As with most federal lands, state fish and game laws apply in national wildlife refuges.

The National Environmental Policy Act has regularly been resorted to by environmentalists who claim that the federal land management agencies have mismanaged water resources under their care. Environmental impact statements showing a consideration of impacts on water resources have been sought against the Forest Service for spraying insecticides, building roads, and allowing clearcutting in national forests, against the Bureau of Land Management for allegedly permitting overgrazing and destructive off-road vehicle use on public lands, and against the Fish and Wildlife Service for permitting sunrise and sunset hunting and the use of lead shot.

Chapter 19
FLOODPLAINS
PROTECTION

Floodplains, the areas adjacent to waterbodies that are periodically required to hold or carry flood flows, are attractive to developers.

> In addition to the obvious aesthetic, recreational, production and transport features of riverfront location, floodplain parcels are often attractively inexpensive, available in larger blocs, and strategically located near developed areas. Moreover, they do not require extensive grading and site preparation. Where governmental control works or other subsidies are available, floodplain investment can produce windfall profits for land speculators.[22]

Approximately 5% of the United States population now lives on floodplains at a time when our society is becoming increasingly aware of the enormous economic and environmental costs of flood control structures.

There is no doubt that "extensive encroachment upon floodplain storage areas will result in significant increases in flood flows and flood hazards."[22] However, the common law can do little to inhibit floodplain development. As pointed out in Part I, no riparian landowner has the right to obstruct the flow of a river if downstream riparians are injured. But in order to win a lawsuit against an upstream riparian, a downstream plaintiff must prove "causation," that is, the damage would not have occurred without the defendant's construction. This is especially difficult to prove in flood damage cases "where the hydrologic and economic injury effects of local obstructions co-mingle with those of myriad upstream users."[22] Moreover, "acts of God" (extraordinary floods) and "sovereign immunity" defenses often preclude recovery. Even if the downstream riparian wins his case, other existing and prospective upstream floodplain developers remain unaffected by the judgment.

THE NATIONAL FLOOD INSURANCE PROGRAM

The major federal statutes enacted to control floodplain development are known as the National Flood Insurance Program (NFIP). Under the NFIP, the federal government provides insurance protection directly and through private insurance companies in flood-prone areas where flood insurance was previously unavailable. In return, communities must adopt zoning ordinances to discourage incompatible development in floodplains and require floodproofing of allowed structures.

The federal agency in charge of the NFIP, now the Federal Emergency Management Agency (FEMA), was first required to define flood hazard zones. In general, these are lands on the 100-year floodplain. The FEMA then had to map ("delineate") these zones in participating communities. While the mapping was in progress, during the "emergency phase" of the program, existing properties and new construction received federally subsidized flood insurance if the community adopted quite minimal standards to restrict floodplain development. The "regular phase" of the program began with the publication of Flood Insurance Rate Maps (FIRMs) delineating zones within the floodplain which were subject to different levels of danger. These FIRMs must have been completed by 1983. After the publication of a FIRM, any new construction within the floodplain must pay full actuarial (true risk) flood insurance rates instead of subsidized rates. However, existing structures continue to receive subsidized flood insurance during the regular phase until they are damaged and require repairs greater than 50% of their market value. The availability of flood insurance during the regular phase is contingent on the community adopting federally approved floodplain regulations that are more stringent than those required during the emergency phase.

Ironically, the NFIP has encouraged floodplain development.[23,24] In the first place, the NFIP has made flood insurance available in high-risk areas for the first time. The subsidized rates offered during the emergency phase have not discouraged people from building in areas subject to floods. Second, a great deal of construction has been undertaken during the emergency phase because subsidized rates are retained for the existing construction once a community enters the regular phase. Third, developers have begun immediate construction during the emergency phase in order to avoid the more stringent floodplain regulations required during the regular phase. Finally, the FEMA has not adequately monitored or enforced the obligation of local communities to police their floodplain regulations.[24] Realistically, communities must participate in NFIP or else forfeit development and tax ratables.

However, once enrolled in NFIP the tendency toward "slippage" (chronic nonenforcement) is strong.

How does the availability of federal disaster aid affect development in floodplains? One commentator argues that disaster aid encourages development.[22] Another believes that the federal disaster assistance program has the "potential to elicit conformity with federal floodplain use standards" because (1) disaster assistance for rehabilitating property is not available in nonparticipating communities; (2) the FEMA grants for reconstructing public facilities and private nonprofit facilities are contingent on construction practices complying with federal standards that emphasize hazard mitigation; and (3) Small Business Administration long-term, low-interest disaster loans for the restoration of homes and businesses will pay for relocating and replacement if the disaster victim is prevented from rebuilding by floodplain regulations, or else will provide funds to meet the requirements of city, county, and state construction codes.[25]

Congress has recognized that federal subsidies, including federal flood insurance, may stimulate development in high-risk areas. The Coastal Barrier Resources Act prohibits all federal financial assistance for development on undeveloped coastal barrier islands. Is this the first step toward a congressional reappraisal of the NFIP? Perhaps its high cost to the taxpayers will finally prove its undoing. Between 1968 and 1981 the NFIP lost $1.5 billion.[24]

STATE FLOODPLAIN REGULATION

Partly because of the NFIP, at least 31 states have adopted programs that regulate fills and structures in the 100-year floodplain:

> These programs either directly regulate fills and structures or set state standards for local regulation of such activities. In general, fills and structures are prohibited in channels and floodway areas, but are permitted in outer "fringe" areas if elevated beyond the highest point reached under 100-year flood conditions or otherwise protected. Although most states rely on maps provided by the NFIP, some have prepared their own maps—for instance, Maryland, New Jersey, and California. Arizona has supplemented its regulatory efforts with acquisition of floodplain areas.[14]

In many of these states, stream encroachment permits are required for construction in nondelineated floodplains.

The major legal problem with floodplain regulations is their vulnerability to "taking" claims. The taking issue has been briefly discussed in previous chapters, but its special importance in floodplains and wetlands cases dictates a more thorough examination here. At some point, a regulation such as a floodplain ordinance will become a taking, requiring the payment of compensation to a burdened landowner, but it is difficult to predict in abstract terms what that point will be. Each case is unique and must be decided on its own facts.

Moreover, courts apply four different tests in determining when a regulation has become a taking. The "diminution of value" test looks to the amount of economic value remaining to a landowner after the imposition of a regulation. If he is not left with substantial economic value, compensation must be paid. For example, it is not a taking for a community to prohibit sand and gravel operations as long as landowners can use their properties for residential or business purposes. Under this test a landowner is not entitled to the highest economic use of his land, but is entitled to reasonable economic uses. The second test, the "private loss-public gain" test, balances diminution of private value against societal gain from the regulation. If the diminution of private value is great, the social need must be compelling for the regulation to be upheld. For example, a fire department may totally destroy a dwelling in order to prevent the spread of fire without having to consider compensating the owner. The third test may be referred to as the "public injury-public benefit" test. Here, it is never a taking when a regulation prevents a public injury, but it is a taking when government uses its regulatory power to confer gratuitous benefits on the public. It would be a taking, for example, to zone private property for use as a public park. The final and most recently formulated test is the "overriding ecological value" test. According to this test, no landowner has the right to develop an ecologically sensitive tract of land, regardless of whether he was aware of its ecological value when the property was purchased.

Floodplain regulations have been struck down under the diminution of value test on the one hand, and upheld under the private loss-public gain and public injury-public benefit tests, on the other. Thus far the overriding ecological value test has been applied only in wetlands cases. The modern trend is away from diminution of value as the sole criterion of regulatory takings. Judges have become more sympathetic to environmental regulation that is reasonably related to a strong public need, despite the effects on individual private landowners. In all likelihood, a reasonable floodplain regulation will be upheld because it is necessary to prevent a grave public danger—flooding.

Chapter 20
WETLANDS PROTECTION

STATE AND FEDERAL PROGRAMS

Because of the profound ecological importance of wetlands and because wetlands frequently transcend municipal boundaries, many states have enacted wetlands protection legislation of one kind or another. Thirteen eastern and Gulf states require permits for the alteration of coastal wetlands. California, Delaware, Maine, New Jersey, and Washington have comprehensive coastal zone protection statutes. In Maine, Michigan, Minnesota, Washington, and Wisconsin, development in the beds of inland waterways and related shoreland development is regulated. Connecticut, Florida, Massachusetts, Michigan, New Hampshire, New York, New Jersey, and Rhode Island require permits for the modification of inland wetlands. State floodplain statutes, environmental impact statement requirements, wild and scenic river systems, and water pollution control statutes often protect wetlands indirectly. Moreover, wetlands can lie within the jurisdictions of regional agencies such as the Adirondack Park Agency in New York and the Pinelands Commission in New Jersey.

Until the early 1970s, landowner plaintiffs were generally successful in their contentions that state and local wetlands regulations as applied to them amounted to unconstitutional takings of their property.[26] But in 1972 the Wisconsin Supreme Court decided the case of *Just* v. *Marinette County*,[27] firmly establishing the "overriding ecological value" test as the takings

142

criterion in wetlands cases. In upholding the Marinette County shoreland zoning ordinance, adopted under a state enabling act, against the constitutional claims of owners who purchased their property before the ordinance was passed, the court said with reference to the plaintiffs' proposed fill of shoreland wetlands:

> An owner of land has no absolute and unlimited right to change the essential natural character of his land so as to use it for a purpose for which it was unsuited in its natural state and which injures the rights of others. . . .
>
> The Justs argue their property has been severely depreciated in value. But this depreciation of value is not based on the use of the land in its natural state but on what the land would be worth if it could be filled and used for the location of a dwelling. While loss of value is to be considered in determining whether a restriction is a constructive taking, value based upon changing the character of the land at the expense of harm to public rights is not an essential factor or controlling.

In other words, the Justs could not lose what they never had; and they never had a right to develop an area of overriding ecological value. Interestingly, the Wisconsin Supreme Court relied in part on the public trust doctrine, which Wisconsin by statute has applied to navigable waters. But courts in other states have followed the *Just* v. *Marinette* reasoning without mentioning the public trust doctrine.

In Part II, several provisions of federal law that might be used to limit unwise development in wetlands were identified. These include, in addition to section 404, section 10 of the Rivers and Harbors Act, the National Environmental Policy Act, the Endangered Species Act, the Wild and Scenic Rivers Act, Executive Order 11990, and the sole source aquifer section of the Safe Drinking Water Act. In addition, point and nonpoint source pollution control under the Clean Water Act can preserve water quality in wetlands.

There are also federal programs oriented toward acquisition of wetlands interests. The Wild and Scenic Rivers Act, the Coastal Zone Management Act, and the Endangered Species Act all authorize funds for land acquisition. Section 73 of the Water Resources Act authorizes the Army Corps of Engineers to purchase land for nonstructural flood control. The Corps can also use dredged spoils to create wetlands at federal cost. The Land and Water Conservation Fund Act provides federal funds for the purchase of natural areas, including wetlands, by the federal government and states. The Water Bank Act establishes a program under which the Department of Agriculture may lease farmlands on a 10-year renewable basis in order to preserve wetlands. The National Estuary Protection Act set up the Wetlands Inventory and provides funding for state acquisitions of estuarine areas. Federal wildlife laws authorizing wetlands acquisition are the Anadromous Fish Conservation Act, the Migratory Bird Conservation Act, the Wetlands Loan Act, the Federal Aid in Wildlife Restoration Act, and the Federal

Aid in Fish Restoration Act. The Migratory Bird Hunting and Conservation Stamps Act authorizes the use of proceeds from duck stamps to fund acquisition of migratory waterfowl habitat. Using these programs, the federal government has protected an estimated 10 million areas of wetlands in the lower 48 states through federal ownership, leases, and easements. About half of this acreage is administered by the U.S. Fish and Wildlife Service and is included in the National Wildlife Refuge System, Waterfowl Production Areas, and Coordination Areas.[28]

The Internal Revenue Service allows tax writeoffs to landowners who donate wetlands to governmental units or conservation groups. Indeed, tax incentives are one of the most successful tools employed by all levels of government to protect wetlands.[29] Among these incentives are property tax abatements and deductions from income, estate, and gift taxes for donations of wetlands.

Ironically, other federal policies have encouraged farmers to clear and drain wetlands. For example, tax writeoffs, commodity price supports, and technical assistance programs have all elevated agricultural productivity over wetland preservation. But a major retreat from this developmental position was taken by Congress in the 1985 farm bill. All federal agricultural subsidies will be denied to "swampbusters"—farmers who cultivate surplus crops on newly converted wetlands.

SECTION 404

Section 404 of the Clean Water Act, which makes it unlawful to discharge dredged or fill material into waters of the United States without a Corps permit, is the major federal regulatory program for preserving inland wetlands. As pointed out in Part II, section 404's geographical coverage is broader than section 10's coverage because section 404 gives the Corps jurisdiction over all "waters of the United States," an area as wide as federal regulatory powers under the Commerce Clause. Thus, in its regulations,[30] the Corps defines "waters of the United States" to include "wetlands," defined as:

> those areas that are inundated or saturated by surface or groundwater at a frequency and duration sufficient to support, and that under normal circumstances do support, a prevalence of vegetation typically adapted for life in saturated soil conditions. Wetlands generally include swamps, marshes, bogs, and similar areas.

The existence of wetlands vegetation is determinative, and it is not necessary that the area be periodically inundated in order to be considered a wetland.

However, section 404 is narrower than section 10 with regard to the scope of activities covered. Section 404 applies only to a "discharge of dredged or fill material," meaning that the material must emanate from a discrete pipe, ditch, channel, or conveyance and be deposited in a wetland. As a result, 404 permits are unncessary for projects involving excavation, drainage, clearing, and flooding of wetlands, or activities on nonwetland areas that damage wetlands, where no fill is deposited in the wetland itself. According to the Office of Technology Assessment of the U.S. Congress, "such activities were responsible for the vast majority of inland wetlands conversions between the mid-1950s and the mid-1970s."[31] Courts have mitigated this problem slightly by holding that construction of a drainage ditch or a levee in a wetland requires a 404 permit, as does the clearing of a wetland's vegetative cover with construction equipment.[32]

Another major difference between section 404 and section 10 is that section 404 is jointly administered by the Corps and the Environmental Protection Agency. The EPA's role in the section 404 process is to (1) promulgate jointly with the Corps "Guidelines for Specification of Disposal Sites for Dredged or Fill Materials"; (2) veto the issuance of permits if necessary to protect the environment; (3) designate geographic areas where no permits may be issued; (4) approve and oversee the delegation of 404 permitting to states; (5) take independent enforcement action against unauthorized dischargers if the Corps does not act; (6) evaluate the impacts from federal construction projects that are exempt from 404; and (7) make final wetlands determinations.

In 1977, Congress exempted from section 404 certain activities that it believed were more appropriately supervised by the states or other federal agencies. These exemptions are (1) normal farming, silviculture, and ranching activities, including "minor drainage"; (2) maintenance and reconstruction of flood control and transportation structures; (3) construction and maintenance of farm or stock ponds or irrigation ditches and the maintenance, excluding the construction, of drainage ditches; (4) construction of temporary sediment basins on construction sites, as long as no fill is placed in waters of the United States; (5) any activity that is being regulated by a state under section 208 of the Clean Water Act (best management practices for nonpoint sources of pollution); (6) construction or maintenance of farm roads, forest roads, or temporary mining roads, where the roads are constructed and maintained with best management practices; and (7) projects planned, financed, and constructed by a federal agency where the environmental impacts of the discharge are addressed in an environmental impact statement submitted to Congress prior to authorization or appropriation of money for the project. The first six exemptions are unavailable if the proposed activities are new water uses that would impair the flow or circulation of waters or reduce their reach. Thus, a farmer is unable to procure an exemption under the normal farming category when he is converting wetlands into dry, arable farmland.

General permits have been one of the most controversial aspects of section 404. As mentioned in Part II, no notification or application is required for specific activities under general permit. Under the latest Corps regulations, all discharges into the following waters are permitted generally: (1) nontidal rivers, streams, and their lakes and impoundments, including adjacent wetlands, that are located above the headwaters; and (2) nontidal waters, including adjacent wetlands, below the headwaters that are not part of a surface tributary system to interstate or navigable waters. "Headwaters" are defined as the point on a nontidal stream above which the average annual flow is less than 5 cubic feet per second. Thus, discharges above the headwaters and discharges below the headwaters into isolated waters are automatically permitted. Moreover, the regulations also grant nationwide general permits to certain identified classes of activities such as discharges of materials for backfill for utility line crossings, materials for bank stabilization to prevent erosion, fill for minor road crossings, and replacement fill.

There are a number of safeguards against abuses of general permits. General permits may not be granted for activities "which cause the loss or substantial adverse modification of 10 acres or more of such waters of the United States, including wetlands." Between 1 and 10 acres, the district engineer must be notified. Moreover, general permits for classes of identified activities must only be issued after public notice, opportunity for comment and public hearing, and public interest review. No general permit is valid if (1) the discharge is located near a public water supply intake; (2) the discharge would destroy an endangered or threatened species or adversely modify its critical habitat; (3) the discharge consists of toxic pollutants in toxic amounts; (4) the fill is not properly maintained and causes erosion; (5) the discharge occurs in a component of the National Wild and Scenic Rivers System; or (6) specified best management practices are not followed. The EPA can veto a general permit for inconsistency with jointly promulgated disposal site guidelines. A state can veto a general permit by denying a water quality certification under section 401 of the Clean Water Act, or can refuse to certify that the activity is consistent with its coastal zone management plan. Moreover, the Corps has discretion to require individual permits for activities that are ordinarily permitted generally. Finally, general permits expire and must be reviewed after five years. However, all these potential safeguards presuppose a high degree of public vigilance. Where a general permit is concerned, the burden of coming forward to show environmental damage is on the public; there is no initial burden on the discharger to show that the environment is not being harmed. And the Corps does little or no compliance monitoring for general permits.[31] How well can the public, for the most part without expert legal or technical assistance, play this watchdog role? This question is especially important where section 404 is concerned because, unlike section 10, citizens can sue to enjoin 404 violations.

If a proposed project does not qualify for a general permit, an application for an individual permit must be filed with the District Engineer (DE). After brief initial review, the DE prepares a public notice of permit application and circulates it to federal, state, and local governments as well as members of the public who have requested notice. Comments must be solicited from the Fish and Wildlife Service, the EPA, and state fish and wildlife agencies. Usually, public comments are accepted for a 30-day period. A citizen may request a public hearing, but granting a hearing is within the DE's discretion. As a rule of thumb, a hearing is held only when there is "significant" public interest in a permit application. The DE then prepares an environmental assessment of the proposed project. If significant environmental impacts are predicted, an environmental impact statement must be prepared and circulated. After this process is completed the DE either grants the permit as requested, grants it subject to conditions, or denies the permit. Only under exceptional circumstances, for example, when the EPA or the Fish and Wildlife Service objects, is a permit "escalated" to the Division Engineer or to Washington.

Applications for 404 permits are subject to the Corps' "public interest review" (see Chapter 11). This is a process of balancing the economic benefits of the proposed project against its environmental detriments. Corps regulations stipulate that the public interest review of a particular proposal must consider the cumulative impacts of discrete development in a wetland area. "Mitigation" is also part of the balance; i.e., the Corps considers project modifications and compensatory mechanisms (e.g., donating comparable wetlands) in determining whether a permit should be granted or denied. The Corps' "upfront" consideration of mitigation measures has been one source of friction between it and agencies such as the EPA and FWS, which would apply mitigation measures only to projects that survive the public interest review.

The EPA's 404(b)(1) guidelines[33] are an integral part of the Corps' public interest review. No permit may be granted "if there is a practicable alternative to the proposed discharge which would have less adverse impact on the aquatic ecosystem." "Practicability" involves consideration of cost, existing technology, and "logistics in light of overall project purposes." Procuring an area not presently owned by the applicant may, under certain circumstances, be a practicable alternative. Where the proposed development is not "water dependent" ("does not require access or proximity to or siting within the [wetland] to fulfill its basic purpose") practicable alternatives not involving wetlands are presumed to be available, unless clearly demonstrated otherwise. Moreover, discharges into wetlands must be shown to be the least damaging practicable alternative.

The EPA and the Corps have had sharp disagreements over interpreting and applying the "practicable alternative" and "water dependency" tests. In one situation the EPA vetoed a Corps permit for a shopping mall on a wetland in Attleboro, Massachusetts. The two agencies could not agree about whether the project was water dependent, whether mitigation should be balanced against adverse environmental impacts, and whether practicable alternative sites existed. The EPA argued that it is the best judge of the meaning of its own guidelines. The Corps responded that the public interest review does not balance economic "goods" and environmental "bads" evenly. Corps regulations contain a presumption in favor of development ("a permit will be granted unless the district engineer determines that it will be contrary to the public interest"). Moreover, the regulations also contain another presumption in favor of projects receiving favorable state determinations, and the Massachusetts Department of Environmental Quality Engineering had issued a state permit for the shopping mall. Both the EPA and the Corps are adamant about their positions; thus, once again— as has occurred regularly during the life of section 404—the courts will have to resolve an important policy issue.

When it was enacted in 1972, section 404 violated three cardinal rules of institutional design: (1) never place two agencies with different agendas in charge of a single program; (2) never place an agency with a particular philosophy in charge of a program based on a conflicting philosophy (or, never place the fox in charge of the chicken coop); and (3) never limit an agency's authority to only a small part of a much larger problem (in this case, to dredge-and-fill activities and not drainage, clearing, flooding, etc.). Given their different orientations, it is not surprising that the EPA and the Corps disagree profoundly about section 404. Given its traditional development role, it is not surprising that the Corps has taken a narrow view of its own jurisdiction, expanded general permits, and "experienced many problems in monitoring permitted activities and enforcing permit conditions."[31] Given the dredge-and-fill focus of section 404, it is not surprising that we are still losing approximately 300,000 acres of mostly inland wetlands per year.[28] And given these deficiencies in the federal program, it is not surprising that more and more states are enacting state wetlands protection statutes. The only pleasantly surprising fact is that the 404 program is apparently saving at least 50,000 acres of wetlands per year.[28]

Chapter 21

WEATHER MODIFICATION

"Weather modification" is defined as "any artificially produced change in the composition, behavior, or dynamics of the atmosphere, when such change is produced with intent to alter the weather."[34] This definition excludes inadvertent weather changes such as those potentially caused by fossil fuel burning, and it implicitly excludes "climate modification" involving relatively long-term or lasting climatic changes. Seeding clouds with silver iodide or other chemicals to increase precipitation, especially to augment winter snowpack, is the most popular weather modification technique. But attempts are also made at hail and lighting suppression, hurricane diversion, and fog dissipation.[35]

As weather modification technology advances, and as intentional climate modification becomes possible, international and national regulation will be necessary because (1) the potential effects of weather modification transcend political boundaries, and (2) as a common good, the atmosphere is not amendable to individual or national ownership. However, there is presently no international law specifically applicable to weather modification, and U.S. legislation is meager. Public Law 92–502, enacted in 1971, requires individuals and nonfederal public entities to report weather modification activities to the Commerce Department's National Oceanographic and Atmospheric Administration (NOAA), which compiles and publishes these reports. The National Weather Modification Policy Act was directed toward developing "a comprehensive and coordinated national policy and a national program of weather modification research and experimentation." Under this

149

act, NOAA is to conduct studies, report to Congress on the status of weather modification technology, and make recommendations as to necessary funding and institutional or legislative changes desirable to facilitate or control weather modification. In other words, federal activity with reference to weather modification has been "preregulatory."

Nevertheless, approximately 30 states have enacted some form of weather modification legislation.[35] These statutes typically contain one or more of the following:

1. A definition of weather modification limited to artificial nucleation of clouds.
2. Establishment of an administrative agency, board, or commission to review and approve weather modification activities.
3. Prerequisites for license and permit application approval, including qualifications for competency and financial responsibility of the applicant, and standards for the method of proposed operation.
4. A requirement that modifiers maintain records of their operations.
5. A requirement for publication of notice in the target area of modification activities.
6. A declaration of legislative intent..

A few states require public hearings or environmental assessments prior to license approval. Some Great Plains states have enacted legislation authorizing the formation of weather control districts by local balloting or by petition. District supervisory boards can tax lands within the district and hire cloud seeders.

As is often the case with new technologies, weather modification has generated a number of lawsuits between private litigants. Thus far at least, litigants have not relied on water law doctrines such as riparianism or prior appropriation.[36] Instead, four tort theories generally pervade these cases: nuisance, trespass, negligence, and strict liability.

Nuisance is a "cause of action" (a set of facts giving rise to an enforceable claim) that allows recovery for an unreasonable interference with a landowner's use and enjoyment of his land. It has frequently been asserted in pollution cases and would be appropriate in lawsuits against weather modifiers. In nuisance cases, judges are entitled to "balance the equities," in other words, to weigh the social value of the defendant's conduct against the plaintiff's loss. Equity balancing might preclude recovery in weather modification cases. For example, in *Slutksy* v. *City of New York*,[37] a resort owner asked for an injunction against cloud-seeding operations near his Catskill Mountain resort by the City of New York, then in the throes of a drought. The court denied the injunction, concluding that the potential benefit of an increased water supply for the city greatly outweighed any loss of business suffered by the resort owner due to lack of sunshine.

Trespass to land is a physical invasion of a landowner's exclusive possession of his land. The trespass may be by visible pieces of matter or by fumes, odors, vibrations, or invisible particles. Unnatural wind or rain caused by weather modification might constitute a trespass and support the award of an injunction and damages. In a Texas case, a cloud-seeding defendant was enjoined from "modifying or attempting to modify any clouds or weather over or in the air space over lands of the [plaintiffs]."[38]

Negligence is the least promising cause of action upon which to predicate recovery against a weather modifier. In order to establish negligence, it must be shown that the defendant violated a duty of due care to the plaintiff by failing to act in a reasonably prudent manner under the circumstances. "One major difficulty with this approach is that so little weather modification activity has been conducted and these controversies have so seldom reached the courtroom, that the establishment of a legal standard of care for a reasonably prudent weather modifier is practically impossible."[35]

Strict liability for an "abnormally dangerous activity" is a more attractive theory. Here, a defendant is liable without fault if he causes damage by using a technology which is unusual in the community and which threatens serious harm even if due care is used. For example, crop dusting is an abnormally dangerous activity in many states. But as with nuisance, courts will balance the equities to determine whether an activity is indeed abnormally dangerous. West Virginia and Pennsylvania, by statute, have imposed strict liability on weather modifiers, while Texas and Utah have enacted statutes requiring proof of negligence as a basis for liability.

The major stumbling block to recovery in weather modification cases, however, is the problem of causation. Under any of these theories, a plaintiff must prove by a preponderance of the evidence that a defendant's weather modification activities were the "legal cause" (direct or "but for" cause) of his damage. The following quotation from an article by Professor Weiss is valuable not only for what it tells us about weather modification lawsuits, but also as an illustration of the differences between "proof" in the legal and scientific communities:

> The most difficult problem in weather-modification litigation remains that of establishing that the weather-modification activity caused the damage in question. For example, plaintiff needs to establish that the increase in rainfall from cloud-seeding caused the flooding which damaged the home or business. Few scientific experts, if any, would be willing to make this statement. At best, we will be able to give a probabilistic statement as to whether or not weather modification could have caused the damage. But not all courts are willing to adjust their procedure to consider a range of statistical probabilities. Usually judges ask only whether it is more probable than not that something occurred or, in a criminal case, whether or not the evidence establishes beyond a reasonable doubt that the defendant committed the crime. Even if we accept the necessity for using probability statistics to establish cause, we will still have uncertainty as to whether or not any given cloud-seeding operation caused

specific damage. For example, if cloud-seeding increased rainfall an average of 25% over a six-year period, rain could have increased 50% or none at all on any single day. The fact that there is a high statistical probability of a 25% increase in rainfall from seeding does not by itself definitely prove that seeding increased the rain by 25% on the day the damage occurred. Even if the rainfall did increase by 25%, the injured party still has to show that this increase in rainfall caused the damage. There may also be difficult questions of how remote the damage can be from the direct impact of the rain and still be eligible for compensation.[36]

Causation problems are further compounded in weather modification cases by the availability of an "act of God" defense or sovereign immunity where a governmental unit attempts to control the weather. There is a ray of hope for plaintiffs, however. Where there has been serious damage to the public health or welfare, courts will sometimes shift the burden of producing evidence as to "non-causation" to the defendant.

Because of these bars to recovery, and because the potential damages from weather modifications are so vast that no insurance company will insure against them, one commentator predicts that as weather modification activities proliferate, the federal government will have to establish an insurance-indemnity fund such as that created for the nuclear industry.[34] But Congress must be cautious lest liability limitations discourage the development and implementation of safer weather modification practices.

Private development of weather modification may also be retarded if firm water rights to the augmented water supplies are unavailable. Under various water allocation theories, water from augmentation practices might belong to the federal government, state governments, modifiers, senior appropriators, riparian owners, or landowners on whose lands the increased precipitation falls. Six of the eight western pure appropriation states have asserted state ownership of atmospheric water.[39] But only Colorado has dealt with the ownership of water produced by weather modification. Under the Colorado statute, a person who augments water supplies following a plan approved by the State Engineer and a water judge has first claim to the waters produced—assuming they can be identified. In other words, Colorado treats such waters as "developed waters."

Chapter 22
ACID PRECIPITATION PREVENTION

Fossil fuel combustion and weather modification are similar in that (1) both activities may cause damage in other states or nations, and (2) because of atmospheric mobility and incomplete scientific knowledge it is nearly impossible to prove that specific damage was caused by a specific source. A rudimentary understanding of how acid precipitation is formed will aid in critically examining the effectiveness of our legal regime for controlling it.

> Sulfur and nitrogen oxides are the compounds most responsible for acid precipitation. Once released into the atmosphere, these gases are converted into sulfates and nitrates through oxidation. The sulfates and nitrates—including sulfuric and nitric acids—are returned to earth in water vapor and small particles of solid material. The presence of sulfuric and nitric acids increases the acidity of the solution and, if not neutralized by other components in the atmosphere, the precipitation falling onto the earth will be acidic. . . . sulfur and nitrogen oxides are byproducts of the burning of fossil fuels.[40]

The dynamics of acid precipitation and sulfate formation are not entirely known.

Acid rain, more accurately described as acid deposition, has been implicated in the destruction of fish and plant life; the contamination of drinking water supplies through the leaching of heavy metals from soil, sediment, and plumbing; the accelerated weathering of buildings, bridges, and monuments; the aggravation of respiratory disease; and the reduction of

153

visibility. Although they have become an issue of national concern, the ecological effects of acid deposition are primarily a problem in northeastern United States and southeastern Canada because of prevailing meteorological conditions and soil characteristics in this area. The major sources of the East's acid precipitation appear to be electrical generating facilities located east of the Mississippi River.

The federal Clean Air Act is the pivotal legal mechanism for controlling air pollution in the United States. This complicated piece of legislation relates to acid precipitation in many ways. But the act as presently administered by the Environmental Protection Agency is incapable of controlling interstate transport of the sulfur oxides, nitrogen oxides, and sulfates that cause acid precipitation.[40]

The Clean Air Act rests on the identification and promulgation by the EPA of health-based National Ambient Air Quality Standards (NAAQS) and state implementation of the national standards in Air Quality Control Regions (AQCRs) within the states through emissions limitations in State Implementation Plans (SIPs). The act does not require uniform national emissions limitations for existing sources of a pollutant if the AQCR is an "attainment area" for that pollutant. Both sulfur and nitrogen oxides are "criteria pollutants" for which NAAQS have been set, but no standard has been set for sulfates. Ironically, most AQCRs in the United States are attainment areas for sulfur and nitrogen oxides. EPA has resisted tightening the NAAQS for these pollutants for the same reason that it has refused to set a sulfate standard—lack of adequate scientific proof on which to base a standard.

The Clean Air Act is oriented to intrastate pollution. States are responsible for attaining and maintaining NAAQS in AQCRs or portions of AQCRs within their borders. In attainment areas, states will protect their own industries by setting emissions limitations as high as possible as long as relevant NAAQS within the state are not exceeded. It is true that the act does contain mechanisms that allow a state affected by interstate pollution to attempt to force the abatement of emissions from a source in another state. However, the downwind state is burdened with the awesome task of proving the source of its acid precipitation. Proof problems restrict the act's interstate pollution provisions to disputes between contiguous states or situations where sources in other states are violating their emissions limitations.[40]

The Clean Air Act's orientation toward reducing state and local pollutant concentrations has actually exacerbated interstate pollution. During the 1970s, states adopted, with the EPA's encouragement, two strategies for dispersing pollutants—tall stacks and "intermittent controls" (managing hours of operation to take advantage of the optimum meteorological conditions for dispersion). Because of court decisions and a clearer understanding of the acid transport problem, intermittent controls are no longer permitted and new tall stacks are discouraged, but numerous "grandfathered" tall stacks remain in use.

A heated debate currently rages over how new controls are to be implemented and who will pay for them. Flue gas desulfurization (scrubbing), coal washing, and fuel switching are expected to produce the quickest reductions in sulfur dioxide emissions. However, it is estimated that the implementation of cleanup plans using these techniques would cost between 3 and 6 billion dollars per year. Alternatively, innovative technology for burning coal cleanly might be developed, but incalculable environmental damage might be done in the meantime.

One way to allocate costs would be to impose the control expenses wholly on the sources of emissions. Under this plan, the electric utility ratepayers in the midwestern states—where rates are generally lower due to lenient air pollution control regulations—would be assessed for the largest portion of the cost. Politicians from the Midwest find this plan "unreasonable" because of potential adverse economic impacts and because the benefits are mainly enjoyed by individuals outside the region. However, politicians in the northeastern states affected by acid precipitation are equally vocal in expressing their lack of sympathy for the ratepayers in states producing two-thirds of the nation's sulfur dioxide emissions.

To lessen or distribute the economic impact of the controls, some legislators have suggested a trust fund approach. Under one version of this proposal, a reduction of 10 million tons of sulfur oxide emissions and 4 million tons of nitrogen oxide emissions would be required over the 48 contiguous states within a decade. (This would constitute a one-third reduction in sulfur oxide emissions.) The 50 most polluting sources would be required to be retrofitted with scrubbers. A consumption tax of one mill (one tenth of a cent) per kilowatt-hour of nonnuclear generated electricity consumed in the 48 contiguous states would finance 90% of the capital costs of control.

This and other proposals are now under serious consideration because of Canada's insistence the United States control its sources of transboundary air pollution.

Chapter 23

CONSERVATION AND REUSE

"Conservation" is a term with a meaning for almost every occasion.[41] "At any time (and for any specific resource use issue) there is an underlying tension between development and preservation views over what constitutes conservation-oriented resource management."[42] Properly defined, water resource conservation should include supply-side structural approaches such as storage reservoirs, in addition to demand-side conservation such as reuse. The legal aspects of transformational conservation methods are discussed in Parts I and II. This chapter concentrates on demand-side nontransformational conservation modes in the agricultural and municipal sectors. Water reuse is given special attention because of its unique public health ramifications. Industrial water conservation through pollution control is addressed in Part IV.

AGRICULTURAL WATER CONSERVATION

"Irrigated agriculture is the largest water user in the United States, accounting for 82% of all flows returned after use."[43] Nearly 90% of all United States irrigation is located in the 17 western states; in some western basins irrigated agriculture makes over 90% of all groundwater and surface water withdrawals. Irrigation efficiency in the West is reportedly under 50%, and technical conservation solutions are available, but "so severe are the existing legal constraints and uncertainties over rights to conserved water, and so strong are the legal protections for those relying on the return flows which would be reduced, that one federal study estimates that only 20% of the total reduction possible through conservation may actually be legally available for new uses."[43]

The prior appropriation system's historic tolerance of waste is examined in Part I. Some of Pring and Tomb's recommendations for change in western water law are:[43]

1. Strict enforcement of conservation practices before granting new diversion rights.
2. Where existing diverters are concerned, state assertion of forfeiture for waste in applications for change in point of diversion, nature of use, or place of use; and enforcement of efficiency standards in general adjudication proceedings.
3. Enforcement of water quality law with a view toward water conservation (better return flow quality is tantamount to more water available for downstream uses).
4. Rejection of "custom" as the primary standard in determining beneficial use and setting duties of water.
5. Recognition of all conserved or salvaged water as becoming the property of the conserver or salvager.
6. Modification of statutory preferences to disallow condemnation by an inefficient "higher use" of an efficient "lower use."
7. Provision of financial incentives to encourage conservation and establishment of "withdrawal charges" scaled to discourage inefficiency.
8. Reform of wasteful Bureau of Reclamation water distribution practices, e.g., increasing artificially low prices of reclamation project water.
9. Institution of "water banking" by a central agency to encourage transfer of inefficiently used water.

Weatherford and Shupe also advocate reforming consumption requirements of interstate water allocation compacts and allowing sales or leases of rights to water from federal projects.[44]

Substantial progress is being made on some of these fronts. California declares water conservation efforts to be "equivalent to a reasonable beneficial use of water" and permits the right to use conserved water to be "sold, leased, exchanged or otherwise transferred" subject to the rights of downstream junior appropriators. But this approach is the exception, not the rule, in the West.[44] California agencies and courts are also beginning to enforce California's constitutional provision that waste of water is not a beneficial use. Colorado, Idaho, and Wyoming prohibit waste, as a nonbeneficial use, in permit and transfer proceedings. Moreover, Oregon and New Mexico treat wasted water as abandoned or forfeited.

Arizona's Groundwater Management Act, summarized in Chapter 5, goes furthest toward mandating conservation practices. For all sectors except agriculture, use of specific conservation technologies is required, and total water-use limitations are imposed. For agriculture, the duty of water has been raised to 70% efficiency. This rate of efficiency is presumed to be attainable by the use of conservation practices such as lined ditches, pump-back systems, land leveling, and efficient application practices.

While most western states allow permanent transfers of water, only seven explicitly allow temporary transfers and only Idaho formally allows statewide water banking.[45] California is lowering legal and institutional barriers to innovative water transfers. The Imperial Irrigation District has been ordered to submit a detailed plan for reducing waste from its irrigation operations. As a result, negotiations are being conducted with the Metropolitan Water District (MWD) for the MWD to fund conservation measures within the IID in return for delivery of the salvaged water to the MWD.

Although progress is being made toward agricultural water conservation in the West, a great deal remains to be done. Most important, little has been done by the federal government or the western states to charge usage rates for water based on actual cost.

MUNICIPAL WATER CONSERVATION

Opportunities for municipal water conservation exist in two main areas: (1) reuse of municipal wastewater, and (2) reduction of demand for water by customers of municipal water supply systems.

It is being recognized throughout the United States that municipal wastewater is an attractive source of reusable water:

> Unlike many agricultural and industrial effluents, municipal wastewater is not subject to salt buildup. The flow of municipal wastewater is relatively consistent over time, unlike highly seasonal agricultural discharges. Municipalities generate vast volumes of wastewater, and sewer systems for collecting and transporting wastewater to a central treatment point already exist in most communities. Finally, because wastewater must be treated to meet pollution control requirements, the effluent produced is already of sufficient quality for some secondary uses. With additional treatment, this effluent can be put to many other uses.[46]

But this additional treatment will probably not be undertaken, because most people are reluctant to accept reclaimed wastewater for drinking or hygienic purposes. Thus, reuse of municipal wastewater means, practically speaking, the use of municipal effluent treated to meet pollution control standards for irrigation, industrial applications, groundwater recharge, and noncontract recreation.

One legal obstacle to reuse of municipal wastewater may be the rights of downstream appropriators in the West to the maintenance of return flows.

The trend in western judicial decisions, however, is to define municipal effluent either as wastewater, developed water, or foreign water. Consequently, western courts have upheld the right of a municipality to discontinue wastewater discharges into a waterbody despite the claims of other appropriators. Furthermore, a California statute grants the owner of a wastewater treatment plant an exclusive right to the treated effluent.

Ironically, municipal water priorities have weakened the incentives to reuse wastewater. Municipalities in both the East and the West have top priority to divert virgin water under statutory preferences, contracts (e.g., with the Bureau of Reclamation), and common law preferences such as "pueblo rights" in California. The conservation disincentive of municipal use priority has led to recommendations that municipal preferences to virgin water be limited to uses that require virgin water.[46] Reused municipal wastewater is also at a competitive disadvantage to diversion of virgin water because the latter has been seriously underpriced, especially in the West. Because it is unlikely that the price of virgin water will ever reflect its true costs, subsidies for municipal reuse will be necessary. The federal Clean Water Act authorizes preferential construction grant funding for wastewater reclamation facilities. However, "faced with limited funding and legislative mandate to abate pollution of the nation's waterways, the EPA has primarily funded traditional pollution control facilities, thus deemphasizing wastewater reclamation and other innovative approaches to water quality enhancement."[46] The state of California has actively assisted the financing of wastewater reclamation projects, but the overly stringent treatment requirements set by the California Department of Health Services have undercut these subsidies.[46]

Even secondary uses of reclaimed municipal wastewater could pose risks to public health if treatment plants malfunction or pretreatment standards go unenforced. Moreover, illicit dumping of hazardous materials into sewer lines is common in some areas. The legal problems facing a municipality considering wastewater reuse are similar to those confronting a weather modifier: insufficient knowledge of potential health risks; extensive potential damage; uncertain liability; and inability to procure private insurance. Once again, if municipal wastewater reuse is to progress in the United States, public insurance-indemnity funds may have to be established to reassure municipalities and compensate victims.

Reduction of demand for water by customers of municipal water supply systems does not encounter such formidable legal hurdles. Demand reduction measures generally fall into three categories: (1) pricing policies, (2) management practices, and (3) pure conservation modes.[47] Pricing strategies include overall rate increases, summer surcharges, and conservation pricing (flat or increasing block rates) instead of decreasing block rates for large users. Leak control and metering are examples of management practices. Pure conservation modes are revised plumbing and building codes, public education, and retrofitting devices on showers and commodes. It seems clear that reasonable demand reduction techniques are well within the legal authority of most municipalities, although private water companies would generally require approvals by state public service commissions.[48] However, in some states, municipalities are prohibited from making "profits" on water sales, thus limiting their flexibility to adjust rates.

Fifteen states—mostly located in the East—have some form of municipal water conservation program using one, two or more of the three categories of demand reduction measures.[49] California and Florida have the most active programs, with Arizona, Massachusetts, New Jersey, North Carolina, and Oklahoma making up a secondary group with somewhat lesser programs. Four states (Pennsylvania, Massachusetts, Oklahoma, and North Carolina) require public water supply systems to have formal conservation programs in order to qualify for state water supply funding.

THE SUPREME COURT AND WATER CONSERVATION

The U.S. Supreme Court's power to equitably apportion the waters of interstate rivers among contending states is examined in Part I. For many years the Court has considered waste and potential conservation as factors in determining equitable apportionments. In a 1982 decision[50] the Court stated:

> Our prior cases clearly establish that equitable apportionment will protect only those rights to water that are "reasonably acquired and applied." Especially in those Western states where water is scarce, "(t)here must be no waste of the 'treasure' of a river. . . . Only diligence and good faith will keep the privilege alive." Thus, wasteful or inefficient uses will not be protected. Similarly, concededly senior water rights will be deemed forfeited or substantially diminished where the rights have not been exercised or asserted with reasonable diligence.
>
> We have invoked equitable apportionment not only to require the reasonably efficient use of water, but also to impose on states an affirmative duty to conserve and augment the water supply of an interstate stream.

It is probable that the Supreme Court's lead will soon be followed by state legislatures, courts, and administrative agencies over a broad spectrum of water rights allocations.

REFERENCES FOR PART III

1. Sax, J. "The Public Trust Doctrine in Natural Resources Law: Effective Judicial Intervention," *Michigan Law Review* 68:471–556 (1970), p. 475.
2. 146 U.S. 387, 13 S. Ct. 110, 36 L. Ed. 1018 (1892).
3. 33 Cal. 3d 419, 189 Cal. Reptr. 346 (1983).
4. Stone, A. B. "Public Rights in Water Uses and Private Rights in Land Adjacent to Water," in *Waters and Water Rights*, R. E. Clark, ed. (7 vols.; Indianapolis, IN: Allen Smith Co., 1967–1978), I, p. 198. The seven-volume treatise is hereafter referred to as "Clark."
5. Meyers, C. J., and A. D. Tarlock. *Water Resource Management*, 2d ed. (Mineola, NY: Foundation Press, 1980), p. 1035.
6. Dewsnup, R. L. "Public Access Rights in Waters and Shorelands," National Water Commission, NTIS PB 205 247 (1971), pp. 12–13.
7. Marvel, C. C. "Public Rights of Recreational Boating, Fishing, Wading, or the Like in Inland Stream the Bed of which is Privately Owned," 6 *American Law Reports* (ALR) 4th (Rochester, NY: Lawyers Cooperative Publishing Co., 1981), pp. 1030–1053.
8. Sax, J. L. and R. H. Abrams, *Legal Control of Water Resources*. (St. Paul, MN: West Publishing Company, 1986), p. 53.
9. Corbridge, J. N. "Surface Rights in Artificial Watercourses," *Natural Resources Journal* 24:887–928 (1984), p. 906.
10. *Anderson* v. *Bell*, 433 So. 2nd 202 (1983).
11. 154 N.W. 2d 473 (1967).
12. 444 U.S. 164, 100 S. Ct. 383, 62 L. Ed. 2d 332 (1979).
13. Hoffman, R. C., and K. Fletcher, *America's Rivers: an Assessment of State River Conservation Programs* (Washington, DC: River Conservation Fund, 1984), p. vi.
14. Kusler, J. "Roles Along The Rivers," *Environment* 27:18 ff. (1985).
15. Cox, W. E. "Protecting Instream Water Uses in the Riparian Doctrine States," in *Water Resources Law* (St. Joseph, MO: American Society of Agricultural Engineers, 1986), p. 125.
16. Tarlock, A. D. "The Recognition of Instream Flow Rights: New Public Western Water Rights," Rocky Mountain Mineral Law Institute 25:24–1 to 24–64 (1979), p. 246.
17. Lamb, B. L., and H. Meshorer, "Comparing Instream Flow Programs: A Report on Current Status," paper presented at a Specialty Conference on Advances in Irrigation and Drainage: Surviving External Pressures, American Society of Civil Engineers, Jackson, Wyoming, July 20–23, 1983. 17 pp.
18. 438 U.S. 696, 98 S. Ct. 3012, 57 L. Ed. 2d 1052 (1978).
19. Ausness, R. "Water Rights, the Public Trust Doctrine, and the Protection of Instream Uses," *University of Illinois Law Review* 1986: 407–437 (1986), pp. 436–437.
20. 608 F. 2d 1250 (1979).

21. "Water Policies for the Future," National Water Commission, U.S. Government Printing Office (1973).

22. Plater, Z. J. B. "The Takings Issue in a Natural Setting: Floodlines and the Police Power," *Texas Law Review* 52(2):201–256 (1974), p. 207.

23. Shaw, S. R. "New Jersey's Freshwater Wetlands—A Vanishing Resource in Need of Protection," Master's Thesis, Rutgers University, (1982).

24. "National Flood Insurance:—Marginal Impact on Flood Plain Development—Administrative Improvements Needed," U.S. General Accounting Office Report CED–82–105 (1982). The title of this GAO Report both belies its conclusion that NFIP "offers a marginal added incentive to development," and disguises its exclusive focus on barrier islands, where development motivations and costs are atypical of floodplains in general.

25. Holmes, B. H. "Federal Participation in Land Use Planning at the Water's Edge—Floodplains and Wetlands," *Natural Resources Lawyer* 13(2):351–410 (1980), p. 361.

26. McGraw, S. M. "State and Local Wetlands Regulation in the Courts: Constitutional Problems on the Wane," *Harvard Environmental Law Review* 1:496–514 (1976), p. 497.

27. 56 Wis. 2d. 7, 201 N.W. 2d. 761.

28. Barnard, W. D., et. al., "Establishing Priorities For Wetland Management," *Water Resources Bulletin* 21:1049–1054 (1985), p. 1051.

29. Kusler, J. A. *Our National Wetland Heritage—A Protection Guidebook* (Washington, D.C.: Environmental Law Institute, 1983), pp. 99–102.

30. 33 CFR pts. 320–330 (1986).

31. OTA, *Wetlands—Their Use and Regulation* (OTA-0-206, 1984).

32. *Avoyelles Sportmen's League, Inc.* v. *Alexander*, 473 F. Supp 525 (W.D. La. 1979) rev'd on other grounds, sub. nom. *Avoyelles Sportmen's League, Inc.* v. *Marsh*, 19 ERC 1841 (CA5, 1983).

33. 40 CFR pt. 230.10.

34. Wood, L. D. "The Status of Weather Modification Activities under United States and International Law," *Natural Resources Lawyer* X(2):367–392 (1977), p. 368.

35. McKenzie, A. G. "Weather Modification: A Review of the Science and the Law," *Environmental Law* 6(2):387–430 (1976), pp. 392–397.

36. Weiss, E. B. "Weather Modification and the Law," in *Pollution and Water Resources, Columbia University Seminar Series*, G. J. Halasi-Kun, ed. (New York: Pergamon Press, 1981), Vol. XIII, pt. 2, p. 20.

37. 197 Misc. 730, 197 N.Y.S. 2d, 238 (1950).

38. *Southwest Weather Research, Inc.* v. *Rounsaville*, 320 S.W. 2d 211 (1958).

39. Danielson, et. al., "Legal System Requirements to Control and Facilitate Water Augmentation in the Western United States," in Ved. P. Nanda, ed. *Water Needs for the Future* (Boulder, CO: Westview Press, 1977), pp. 294–295.

40. Gallogly, M. R. "Acid Precipitation: Can the Clean Air Act Handle It? *Boston College Environmental Affairs Law Review* 9(3):687–744 (1981–1982), pp. 688–690.

41. Herfindahl, O. C. "What Is Conservation?" in *Three Studies in Mineral Economics* (Washington, D.C.: Resources for the Future, 1961), pp. 1–12.

42. Shabman, L. "Discussion," *Water Resources Bulletin* 18(2):345–347 (1982), p. 346.

43. Pring, G. W., and K. A. Tomb. "License to Waste: Legal Barriers to Conservation and Efficient Use of Water in the West," *Rocky Mountain Mineral Law Institute* 25:25–1 to 25–67 (1979), p. 254.
44. Weatherford, G. D., and S. J. Shupe, "Reallocating Water in the West," *Journal of the American Water Works Association* 1986:63–71 (October, 1986), p. 70.
45. Blackwelder, B., and P. Carlson, *Survey of Water Conservation Programs in the Fifty States* (Washington, D.C.: Environmental Policy Institute, 1982), p. 98.
46. Brown, E. C., and N. Weinstock. "Legal Issues in Implementing Water Reuse in California," *Ecology Law Quarterly* 9(2):243–294 (1981), p. 248.
47. Moomaw, R. L., and L. Warner. "The Adoption of Municipal Water Conservation: An Unlikely Event?" *Water Resources Bulletin* 17(6):1029–1034 (1981), p. 1030.
48. Baram, M. S. "Legal, Economic and Institutional Barriers to Water Reuse in Northern New England," U.S. Department of the Interior Report OWRT/14–34–0001–9424 (1980), pp. 87–96.
49. Sawyer, S. W. "State Water Conservation Strategies And Activities," Water Resources Bulletin 20:679–685 (1984), p. 681.
50. *Colorado* v. *New Mexico*, 459 U.S. 176, 103 S. Ct. 539, 74 L. Ed. 2d 348 (1982).

PART IV
Water Treatment and Land Use

Water treatment and restrictions on land development are the primary approaches to protecting, restoring, and improving water quality. "Water treatment" is defined here as subjecting water to a cleansing agent or action, or segregating it from contact with other waters. This can take place at any of three points: (1) before wastewater is or would be discharged to a water-body or groundwater; (2) in the process of removing pollutants that presently exist in waterbodies; and (3) after withdrawal for uses such as drinking water. Land development prohibitions are generally imposed only where water treatment cannot protect high-quality waters.

As the following quotation from Leonard B. Dworsky points out, water quality law has evolved independently of other aspects of water law:

> Historically, the strategy for managing water resources is to capture and store waters in order to have them available for use when and where they are needed. Lakes and simple embankments across flowing streams are centuries-old means for accomplishing this. Only within the last hundred years has a new component, based on the scientific concepts of quality arising out of the work of Pasteur and the other early bacteriologists, been added to the age-old strategy. Today the strategy is to make water available when needed, where needed, and of the right quality. For nearly three-quarters of a century, science, technology, and public policy have been moving in the direction of improving society's ability to manage this new component, quality. It has turned out to be much more difficult than the task of managing quantities of water, even flood waters, because water quality or pollution control is involved with so many uses of water by so many people. Although the political and technical

skills for dealing with this issue have been improving, further sophistication is needed before this important problem can be dealt with adequately. An important part of this problem is finding more effective ways to integrate water pollution control into the overall water resources planning and development task.[1]

One goal of modern water law is to facilitate comprehensive water resource management by creating a new synthesis from disintegrated fragments of prior water law.

Chapter 24

THE CLEAN
WATER ACT:
INTRODUCTION

This long and complicated statute[2] combines two approaches to water pollution control: a water quality-based approach and a technology-based approach. In the context of the CWA both are "regulatory" strategies; that is, they entail administrative permitting of allowable discharges and enforcement against permit violators. Nonregulatory "market" mechanisms, such as "effluent charges" and "pollution rights" auctions, have been proposed,[3] but have not been adopted by Congress.

Water quality standards are the focus of the water quality-based approach to pollution control. A "water quality standard" has two parts: (1) a designation of the desired use for a given body of water, and (2) the water quality criteria appropriate for that use. "Water quality criteria" are specific levels of water quality that, if not exceeded, are expected to render a body of water suitable for its designated use. Criteria may be expressed as either a number, a narrative, or both. An example of a numerical criterion is a dissolved oxygen (DO) criterion, which for swimming and nontrout fishing might set a level of 5.0 milligrams of DO per liter of water as a 24-hour average, with no less than 4.0 milligrams per liter at any time. For a "parameter" (aspect of water quality) such as taste and odor, the criterion might be expressed narratively: "None offensive to humans or which would produce offensive tastes and/or odors in water supplies and biota used for human consumption. None which would render the waters unfit for the designated uses."

Water quality standards are measures of "ambient" (instream) water quality. But regulatory agencies frequently use a laboratory surrogate for water quality standards, variously called "biomonitoring," "bioassay," or "whole effluent testing" (hereafter "bioassay"). A bioassay is conducted by subjecting organisms representative of the aquatic ecosystem to a particular effluent and quantifying the concentration of the effluent necessary to kill 50% of the organisms within a 96-hour period. Water quality standards are then set at a percentage of this lethal concentration.

For permitting and enforcement purposes, water quality standards must be translated into "water quality-based effluent limitations" for particular dischargers. "Effluent limitations" are restrictions on quantities, rates, and concentrations in wastewater discharges measured at the discharger's outfall pipe. A water quality-based effluent limitation is established using a water quality model, a mathematical technique for predicting the effect of discharges on ambient water quality. Based on the model, dischargers are given "wasteload allocations" specifying each discharger's share of the "Total Maximum Daily Load" (TMDL) of pollutants that can be discharged into a waterbody without violating the water quality standard. A discharger's wasteload allocation is expressed as a water quality-based effluent limitation.

Before 1972, American water pollution control policy was predicated on the water quality-based approach, and it was a failure.[4] First, there is not yet an adequate scientific basis for tying water quality criteria to designated uses. The following passage is taken from a U.S. General Accounting Office (GAO) report criticizing the Environmental Protection Agency's water quality criteria as being too strict, but the same argument could be used to attack other criteria as overly loose:

> DO concentrations are important in gauging water quality. In fact, DO has been called "Probably the single most important water quality parameter in fisheries management." Therefore, a complete and thorough scientific basis for the recommended DO criterion would be expected, but cannot be found. Scientists disagree considerably on how much DO fish need. Most species of adult fish (including brook trout) can survive at very low DO concentrations. Minimum tolerable levels reported by some investigators are several times greater than those reported by others for the same fish species, tested at about the same temperatures. Many apparent contradictions also exist about the effects of DO levels on hatching of fish eggs and growth of many young fish.[5]

Moreover, general criteria are not appropriate to all waters at all times of the year, nor are they sensitive to differences in intensity, frequency, duration, and extent of violations.

Second, assigning wasteload allocations to dischargers based on mathematical models is still an uncertain enterprise because of the relatively primitive nature of even the most advanced water quality models. According to the GAO, "It is very difficult to model complex natural processes found in many water bodies because so little is known about them.''[5] Most stream models concentrate on DO, ignoring other components such as toxic substances and ''nutrients'' (nitrogen and phosphorus). DO models are themselves imprecise because natural processes such as stream reaeration are not fully understood, ''nonpoint source'' (runoff) loads are neglected, measurement errors are frequent, mistaken assumptions are used in the absence of adequate water quality data, and many models are not properly verified.

Modeling and wasteload allocation difficulties are compounded by the concept of ''mixing zones,'' a traditional part of water quality standards. A mixing zone is an area around a discharge point in which a discharger is permitted to mix its wastes without liability for violating water quality standards. For example, current New Jersey regulations define the boundaries of nonthermal mixing zones as follows:

> The total area and volume of water assigned to nonthermal mixing zones shall be limited to that which will not interfere with biological communities or populations of important species to a degree which is damaging to the ecosystem or which diminishes other beneficial uses disproportionately. Furthermore, mortality of aquatic life shall not occur within the nonthermal mixing zone.

This question-begging definition defies rational application. Its absurdity is typified by the requirement that ''fish corridors'' be provided where, because of multiple dischargers located on a certain stretch, mixing zones tend to encroach on one another.

Third, even if waterbodies could be modeled precisely there would be thorny problems of distributional equity in attempting to allocate wasteloads.[6] Should downstream dischargers receive lower TMDLs because the water is dirtier? If a percentage allocation is utilized, should large and small firms be treated alike? Should waterbodies be zoned for wasteload allocation purposes? If so, where should zonal lines be drawn? What allowances should be made for future growth?

Fourth, the variability of water quality-based effluent limitations is also a major obstacle to enforcement. Identical dischargers located on different waterbodies or even on different segments of the same stream are allocated different TMDLs because of variations in receiving water quality. Enforcement efforts are countered by ''forum shopping''—discharger threats to relocate to states or areas with cleaner water. These threats gain credibility because the firm is placed at a competitive disadvantage simply by virtue of its geographical location. Pressures on pollution control officials to compromise at the expense of water quality are extreme where enforcement might lead to plant closings and consequent economic dislocation.

Fifth, because water quality-based effluent limitations are so difficult to set and enforce, there are many waterbodies or segments of waterbodies where state agencies have not established them, especially for toxic pollutants. In these areas, water quality standards are virtually unenforceable.

When a violation occurs, enforcement officials must ask questions such as, Who did it? When was it done? Was it one polluter or were there a number of them? Was the violation caused by a factory? A municipal sewage treatment plant? A farmer applying pesticides? A "midnight dump" of hazardous wastes? A passing vessel? And if a likely source is identified, how can the violation be conclusively traced to it?

Sixth, many people consider the water quality-based approach to be morally intolerable. It assumes that, to some extent, "pollution is the price of progress." It does not recognize a public right to clean water, but strives to use as much of a waterbody's "assimilative capacity" as is consistent with other social goals.

Finally, our latest refinement on water quality standards—the bioassay—is a highly useful tool for determining the toxicity of mixed waste streams, but it has limitations with regard to water pollution control in general. Bioassays are insensitive to long-term effects, bioaccumulative effects, and synergistic or antagonistic effects of multiple discharges.

In 1972 Congress, disenchanted with the water quality-based approach, redirected America's water pollution control program. The Clean Water Act of 1972 retained water quality standards and wasteload allocations, but only as a temporary, secondary line of defense. Its principal control mechanism is uniform national technology-based effluent limitations, progressively tightened until a "zero-discharge" goal is reached. "Such direct restrictions on discharges facilitate enforcement by making it unnecessary to work backward from an over-polluted body of water to determine which . . . sources are responsible and which must be abated." They focus on "the preventable causes of water pollution" rather than its "tolerable effects."[7] The act's technology-based effluent limitation approach has significantly reduced or prevented water pollution in the United States, although much remains to be done. Debate about the act now involves whether further water quality improvement is worth the cost of installing expensive control technology, and whether these more sophisticated control mechanisms are necessary to restore or maintain desired waterbody uses. There is a movement toward reinstating the water quality-based approach, including an economic component, above the uniform level of technological control already attained. In short, the tension between proponents of the water quality-based and technology-based approaches remains—and will probably continue to remain—unresolved.

The following detailed discussion of the CWA amplifies these introductory remarks. Emphasis is placed on the act's three major water pollution control modes: the construction grants program for reducing municipal discharges, the National Pollutant Discharge Elimination System (NPDES) permit program for control of point source discharges, and water quality management planning for nonpoint source control.

Chapter 25
THE CLEAN WATER ACT:*
GOALS AND POLICIES

Pollution is not a fact of nature but a human value judgment. Society decides what is clean and what is polluted. Prior to 1972, American law implicitly defined "pollution" as the inability to use a waterbody for a particular purpose because of water quality deficiencies. Waterbodies were thought to possess "assimilative capacities" which could be used to decompose or dilute wastes while still maintaining desired uses. Wasteload allocations represented allocations of assimilative capacities, and water quality standards represented desired uses. Pollution meant overloading a waterbody with wastes and exhausting its assimilative capacity, precluding desired uses (violating water quality standards). Before 1972, desired uses were set by individual states, which classified navigable waters in categories ranging from Class A (swimming) to Class D (agricultural and industrial use). If a state was satisfied that a particular river was esthetically tolerable and fit for navigation, the law did not afford relief unless the river stank or corroded hulls. One river, the Cuyahoga in Ohio, was not considered legally objectionable until it caught fire, because the state-designated use of that river was waste disposal. Moreover, some states established weak water quality

---

*The "Clean Water Act" (CWA) is congressional shorthand for the "Federal Water Pollution Control Act." All references to the Clean Water Act are to the original congressional section numbers, not those found in the United States Code. For example, section 101 of the act is officially listed as 33 U.S.C. section 1251. Unofficial section numbers are used here because these are the numbers known by water resources professionals, because the copies of the act distributed by EPA were based on congressional prints, and because the congressional section numbers are easier to remember.

standards in order to retain resident industries and attract new ones.

In summary, water pollution control law prior to 1972 gave a right to discharge until waters were polluted, meaning until state-set water quality standards were violated. The public had a right to water only as clean as the state dictated in its water quality standards. Pollution was defined as an excessive discharge—excessive in the sense that it disqualified the water-body for its intended purpose.

One key to understanding the CWA of 1972 is to recognize that it redefined "pollution." Section 502 of the CWA defines "pollution" as "the man-made or man-induced alteration of the chemical, physical, biological, and radiological integrity of water." What is this "integrity of water" that pollution disrupts? The CWA does not, as some of its critics suggest, attempt to recreate a preindustrial American utopia, a dehumanized state of primeval purity. Instead, the CWA's legislative history[8] makes it clear that what Congress meant by "integrity" of a waterbody was its ecological stability, its ability to respond to change by maintaining ecological diversity. The existence of fish and wildlife in and around a waterbody and its capacity to sustain human recreation are indicators of ecological stability. Moreover, a waterbody's integrity must be viewed in the context of unalterable realities intrinsic to our highly industrialized and urbanized society.

With the CWA's definition of "pollution" in mind, its goals become more intelligible. Section 101 declares:

(a) The objective of this Act is to restore and maintain the chemical, physical, and biological integrity of the Nation's waters. In order to achieve this objective it is hereby declared that, consistent with the provisions of this Act— (1) it is the national goal that the discharge of pollutants into the navigable waters be eliminated by 1985; (2) it is the national goal that wherever attainable, an interim goal of water quality which provides for the protection and propagation of fish, shellfish, and wildlife, and provides for recreation in and on the water be achieved by July 1, 1983; (3) it is the national policy that the discharge of toxic pollutants in toxic amounts be prohibited; (4) it is the national policy that Federal financial assistance be provided to construct publicly owned waste treatment works; (5) it is the national policy that areawide waste treatment management planning processes be developed and implemented to assure adequate control of sources of pollutants in each State; (6) it is the national policy that a major research and demonstration effort be made to develop technology necessary to eliminate the discharge of pollutants into the navigable waters, waters of the contiguous zone, and the oceans, and (7) it is the national policy that programs for the control of nonpoint sources of pollution be developed and implemented in an expeditious manner so as to enable the goals of this Act to be met through the control of both point and nonpoint sources of pollution.

These goals and policies are the CWA's strategies for accomplishing its overall objective—"to restore and maintain the . . . integrity of the Nation's waters"; in other words, to end water pollution.

When interpreting a statute, it is important to distinguish between goals and policies, on the one hand, and enforceable requirements on the other. "Zero discharge" is only a long-term goal. It reminds us that the most efficient way to end pollution is to eliminate residuals—to progressively tighten treatment requirements until waste minimization, recycling and reuse become desirable alternatives. Whatever their motives, those who criticize the zero discharge goal as illusory or impractical are missing the point. This goal has proved useful in mobilizing public effort and interpreting ambiguities in the CWA's language. Moreover, zero discharge has already been attained by major categories of American industry. The CWA should not be condemned as a failure because the 1985 zero discharge goal has not yet been achieved by all dischargers. Goals are inherently flexible, and even enforceable requirements can be modified by Congress if necessary. It has been learned since 1972 that water pollution control is more complicated and expensive than was first anticipated. Thus, it is no disgrace to postpone fulfilling some of our goals and to "fine tune" certain enforceable requirements. In fact, Congress should be proud of having enacted such a dynamic and durable statute.

But why is zero discharge necessary to maintain the integrity of water? Cannot some wastes be discharged without jeopardizing the ecological stability of a waterbody? Theoretically the answer is yes, but in practice the water quality-based approach has proved to be unenforceable. So we have created a presumption that any measurable discharge of any substance into a waterbody is inconsistent with the integrity of its water. We are seeking zero discharge because it is impractical to seek anything less.

Congress has modified the zero discharge goal where it is unrealistic. "Publicly-Owned Treatment Works" (POTWs) are responsible for implementing "secondary treatment," not complete removal (see Chapter 28). Nonpoint sources are only responsible for implementing "Best Management Practices" (BMPs), which take economic and technical feasibility into account (see Chapter 34). Furthermore, water quality-based variances have been established for heat discharges, POTW discharges into marine waters, and discharges of nonconventional/nontoxic pollutants (see Chapter 27). Unlike some inflexible presumptions, the zero discharge presumption has been a rebuttable one when it has come to translating this goal into enforceable requirements.

The CWA's second goal—referred to as the "fishable/swimmable" goal—is both an interim goal and, as will be seen, an enforceable requirement. But it is qualified by the "wherever attainable" language. All waters of the United States must be at least suitable for fishing and swimming, wherever these uses are attainable. States are no longer authorized to designate waterbodies for uses below these minimum standards of cleanliness. No longer is pollution looked upon as an inevitable accompaniment to economic progress. The public is entitled to fishable/swimmable waters, wherever attainable, on the way to zero discharge.

Chapter 26
THE CLEAN WATER ACT: ANTIDEGRADATION AND ATTAINABILITY

ANTIDEGRADATION

Section 101 declares a policy to "restore and maintain" clean water, and "antidegradation" requirements have traditionally been a prominent feature of water pollution control law. Our national goal is not only to clean up dirty waterbodies but also to forestall degradation of presently high-quality waters. Federal regulations promulgated on November 8, 1983, require each state to adopt a statewide antidegradation policy and identify the methods of implementing it.[9] Antidegradation applies whenever a waterbody is better than fishable/swimmable—for example, if trout can propagate there. The state antidegradation policy and implementation methods must safeguard existing water uses such as trout propagation. But if existing water quality is better than needed for existing water uses, a state may, after public participation, choose to allow lower water quality where "necessary to accommodate important economic or social development." Where degradation is allowed the "bottom line," once again, is existing water uses. No degradation at all is permitted in so-called "Outstanding National Resource Waters," such as waters in parks, wildlife refuges, "and waters of exceptional recreational and ecological significance."

Most states have adopted antidegradation policies, but few have meaningful implementation measures. In the first place, the enforceable requirements of the CWA are directed at "point sources"; for example, industries and POTWs. For various reasons, point sources generally do not locate on the headwater streams that are covered by antidegradation. Upland streams are typically degraded by runoff from "nonpoint sources" such as mining, forestry, and residential real estate development. The CWA does not regulate nonpoint sources of pollution. In the second place, states have traditionally been reluctant to regulate these nonpoint sources, especially residential development which is seen as a local concern. And the CWA does not compel states to regulate nonpoint sources. Thus, state antidegradation policies tend to be abstract, unenforceable declarations.

Vermont and Florida have been exceptions to the general failure of state antidegradation policies. The Vermont legislature has prohibited all discharges above 2500 feet elevation and into those streams rated Class A. Florida has adopted strong measures to protect Outstanding National Resource Waters like the Everglades.

ATTAINABILITY

In contrast to antidegradation, "attainability" applies where a waterbody is worse than fishable/swimmable. It has been seen that the CWA establishes a minimum fishable/swimmable use for waterbodies "wherever attainable." But what does this important phrase mean? In the first place, a use is attainable if it is already being attained. Thus, where existing water quality standards specify designated uses less than those that are presently being attained, the state must revise its standards to reflect the uses actually being attained. Second, a use is attainable if it can be achieved by the imposition of technology-based effluent limitations for point sources "and cost-effective and reasonable best management practices for nonpoint source control." Third, a use is deemed attainable unless the state, after a "use attainability analysis," can demonstrate that attaining the fishable/swimmable use is not feasible because:

> (1) Naturally occurring pollutant concentrations prevent the attainment of the use; or (2) Natural, ephemeral, intermittent or low flow conditions or water levels prevent the attainment of the use, unless these conditions may be compensated for by the discharge of sufficient volume of effluent discharges without violating State water conservation requirements to enable uses to be met; or (3) Human caused conditions or sources of pollution prevent the attainment of the use and cannot be remedied or would cause more environmental damage to correct than to leave in place; or (4) Dams, diversions or other types of hydrologic modifications preclude the attainment of the use, and it is not feasible to restore the water body to its original condition or to operate such modification in a way that would result in the attainment of the use; or (5) Physical conditions related to the natural features of the water body, such as the lack

of a proper substrate, cover, flow, depth, pools, riffles, and the like, unrelated to water quality, preclude attainment of aquatic life protection uses; or (6) Controls more stringent than [the CWA's technology-based effluent limitations] would result in substantial and widespread economic and social impact.[9]

The vagueness of these criteria makes it probable that they will be, as their predecessors have been, politically and legally controversial. Serious political and economic consequences depend on how they are interpreted. For example, how much growth will be permitted at the expense of water quality? In water-short areas, what tradeoffs will be permitted between diversions and impaired quality? When is restoration of more natural water quality feasible? These questions reflect the tensions inherent in the concept of "attainability."

Chapter 27

THE CLEAN WATER ACT: INDUSTRIAL POINT SOURCE DISCHARGERS

In the CWA the term "discharge of a pollutant" means "any addition of any pollutant to navigable waters from any point source." As pointed out in Part II, "navigable waters" is generously defined by the CWA as "waters of the United States" including the territorial seas within the three-mile limit. In turn, courts have interpreted "waters of the United States" to include wetlands, drainage ditches, mosquito canals, and intermittent streams. The CWA is silent about whether "waters of the United States" includes groundwater. Different courts have held both ways on the matter. The Environmental Protection Agency, however, has acted as though the CWA's point source discharge requirements do not apply to groundwater discharges.

A "point source" is "any discernible, confined and discrete conveyance, including but not limited to any pipe, ditch, channel, tunnel, conduit, well, discrete fissure, container, rolling stock, concentrated animal feeding operation, or vessel or other floating craft. . . . This term does not include return flows from irrigated agriculture." "Point source" is to be liberally construed, and has been held to include earth-moving equipment in a wetland and ponded mine drainage that is allowed to erode a channel to a waterbody. On the other hand, EPA has discretion to treat point sources as "nonpoint sources" (sources other than point sources) for regulatory purposes. For example, dams are point sources capable of causing diminution in downriver water quality, but EPA, with court approval, treats dams as if they were nonpoint sources. A 1987 CWA amendment includes a landfill leachate collection system as a point source, but excludes an agricultural stormwater discharge.

Point sources are either industrial or municipal (POTWs). Industries are either "direct dischargers" to waterbodies or "indirect dischargers" into POTWs which then discharge to waterbodies. This chapter deals with industrial point source direct dischargers.

The CWA's approach to regulating industrial point source dischargers is to impose progressively stricter technology-based effluent limitations on categories of point sources in pursuit of a zero-discharge goal. Where imposition of technology-based limitations will not produce fishable/swimmable waters where attainable, more stringent water quality-based limitations must be applied.

PHASE I

When the CWA was enacted in 1972, it contained two phases of technology-based limitations. In Phase I existing industrial point sources were required to apply the best practicable control technology currently available (BPT), as defined in EPA regulations, by July 1, 1977. During Phase II, existing industrial point sources were to meet effluent limitations based on best available technology economically achievable (BAT), or, if feasible, zero discharge, by July 1, 1983. New sources were expected to immediately comply with strict standards of performance based on best available demonstrated control technology (BACT), a standard comparable to BAT for existing sources, or zero discharge if practicable. However, having met the applicable standard of performance, the new source could not be required to meet stricter technology-based standards for 10 years or the plant's amortization period, whichever ended first.

In setting effluent limitations based on BPT, Congress instructed the EPA to consider

> the total cost of application of technology in relation to the effluent reduction benefits to be achieved from such application, and . . . also take into account the age of equipment and facilities involved, the process employed, the engineering aspects of the application of various types of control techniques, process changes, nonwater quality environmental impact (including energy requirements), and such other factors as the Administrator deems appropriate.

Legislative history indicates that BPT was seen primarily as available "end-of-pipe" treatment, and the EPA's definition of BPT would be acceptable as long as control costs and economic impacts were not "wholly out of proportion" to water quality benefits. In contrast, BAT was thought of primarily as in-plant process changes that had been or were capable of being achieved. Compliance costs were considered in setting BAT, but no cost-benefit analysis was necessary as with BPT. Congress realized that some businesses would be forced to cut back production or even close down as a result of their inability to afford the costs of complying with these technology-based standards.

The EPA's strategy in setting BPT effluent limitations was to divide industries into categories based on products manufactured and subcategories based on processes utilized. For example, the asbestos manufacturing category contained asbestos-cement sheet and asbestos paper subcategories. Based on an average of the performance of the three cleanest "exemplary plants" in each subcategory, the EPA promulgated regulations containing effluent limitations for relevant parameters set out in terms of maximum daily and monthly averages per unit of production or volume of wastewater. These "single number" effluent limitations were uniform for existing plants in a particular subcategory wherever located. However, a discharger that objected to inclusion within a certain subcategory was entitled to a variance if it proved that "factors relating to the equipment or facilities involved, the process applied, or other such factors . . . are fundamentally different from the factors considered in the establishment of the guidelines." This is known as the "Fundamentally Different Factors" (FDF) variance. As interpreted by the courts and codified by the 1987 CWA amendments, a determination of whether a discharger is entitled to an FDF variance cannot be based on abnormal compliance costs (independent of other eligible factors) or on receiving water quality. Nor can a plant be fundamentally different because of impurities contained in its intake water, because effluent limitations are always set on a "net" basis. Also, effluent cannot be diluted in order to meet effluent limitations. An FDF variance is simply an acknowledgment that the discharger has been placed in the wrong subcategory.

A discharger receiving an FDF variance is either placed in a more suitable subcategory or else becomes a "subcategory of one." In the latter situation, the technology-based effluent limitation is ascertained by best professional judgment (BPJ)—a catchall standard that is resorted to whenever technology-based effluent limitations are unavailable.

During the early and middle 1970s, regulators and general public alike perceived water pollution to be a function of the so-called "conventional pollutants" (i.e., biochemical oxygen demand, total suspended solids, fecal coliform, pH, and oil and grease). Thus, the EPA's original BPT-based effluent limitations were comparatively strict with regard to these parameters. Even before the United States Supreme Court upheld the EPA's categorical approach to setting technology-based effluent limitations,[10] industry had begun to comply with its Phase I requirements. By all acounts, industry has removed over 70% of conventional pollutants from its waste streams. This is a tribute to Congress, the EPA, industry, and the American public. It should be remembered that the CWA, and its technology-based approach, has been a resounding success—in terms of its expectations when enacted—with regard to industrial point source pollution.

The extent of toxic pollutant contamination of our waterbodies was not widely recognized before the middle to late 1970s. As a result, little progress was made during Phase I in controlling "toxic pollutants," defined in the CWA as pollutants or combinations of pollutants which "upon exposure, ingestion, inhalation, or assimilation into any organism, either directly from the environment or indirectly by ingestion through food chains, will . . . cause death, disease, behavioral abnormalities, cancer, genetic mutations, physiological malfunctions (including malfunctions in reproduction) or physical deformations, in such organisms or their offspring."

PHASE II

In 1977 Congress decided that, in light of the progress already made, the cost of moving to BAT for conventionals was too great. A modified requirement, called best conventional pollutant control technology (BCT), was to underlie the development of further technology-based effluent limitations for dischargers of conventionals, and compliance was to be achieved by July 1, 1984. BCT includes two cost tests: (1) a comparison between the costs of reducing discharges of conventionals and the resultant water quality benefits; and (2) a comparison between industrial and municipal treatment costs. Congressional supporters of BCT felt that it would most often fall between BPT and BAT.

The EPA has found the BCT cost tests difficult to apply. In fact, its first set of BCT regulations did not survive court challenge. The EPA responded by promulgating regulations for each industrial subcategory that, in almost all cases, equate BCT with BPT. In the 1987 CWA amendments, Congress extended the BCT compliance dates to three years from the date of promulgation of applicable effluent limitations.

Oriented as it was toward conventional pollutants, the 1972 CWA incorporated inconsistent strategies for controlling toxics. On the one hand, dischargers of toxics were subject to phased technology-based effluent limitations for industrial categories. On the other hand, there was an alternative pollutant-by-pollutant approach requiring EPA to (1) publish a list of toxic pollutants; (2) set effluent standards for these pollutants based not on technology or water quality but on toxicity, persistence, degradability, and ecosystem effects; (3) set a margin of safety; (4) hold lengthy hearings; and (5) enforce compliance with effluent standards within one year of promulgation. Confused by these inconsistent mandates and preoccupied with conventional pollutants and POTW pollution, the EPA virtually ignored industrial dischargers of toxics until numerous environmentalist lawsuits forced EPA into a court-approved settlement. The famous "Consent Decree" of 1976 confirmed the EPA's ultimate decision to regulate toxic pollutants primarily by requiring categories of dischargers to comply with effluent limitations based on BAT. The Consent Decree also established timetables for EPA to promulgate effluent limitations for 23 industrial categories (approximately 500 subcategories) covering 65 families of compounds, which the EPA has broken down into approximately 130 "priority pollutants." Because of the complexity of establishing technology-based effluent limitations for toxics and because of the frequently open hostility of the Reagan Administration, the EPA has taken about a decade to promulgate the stipulated regulations. Congress codified the Consent Decree in 1977, and in 1987 extended compliance dates from July 1, 1984 to three years from promulgation. But in emergency situations the EPA can require immediate cessation of discharges or up to one-year compliance periods.

In addition to conventional and toxic pollutants, Congress, in 1977, created a third class of pollutants called "nonconventional/non toxic" pollutants. Ammonia, chlorine, color, iron, and total phenols are statutorily placed in this category and the EPA may list—as it also may with toxics—additional pollutants on its own initiative or in response to petitions. Dischargers of nonconventional/nontoxic pollutants must meet BAT-based effluent limitations within three years of promulgation. But, unlike dischargers of conventionals and toxics, dischargers of nonconventionals/nontoxics have access to variances based on economics and receiving water quality. First, there is a potential modification that "will represent the maximum use of technology within the economic capability of the owner or operator" and "will result in reasonable further progress toward the elimination of the discharge of pollutants." A second variance provision allows the EPA, with state consent, to impose a substitute effluent limitation which will: (1) be at least as strict as one based on BPT; (2) not result in additional requirements being placed on another point source; and (3) not interfere with the attainment or maintenance of fishable/swimmable waters or endanger human health.

HEAT DISCHARGERS

Heat discharged into water is a pollutant, and dischargers of heat must meet effluent limitations based on BAT by the established compliance date. Nevertheless, under subsection 316(a) of the CWA, a variance based on receiving water quality is available to dischargers of heat. Whenever the discharger can demonstrate

> to the satisfaction of the Administrator (or, if appropriate, the State) that any effluent limitation proposed for the control of the thermal component of any discharge from such source will require effluent limitations more stringent than necessary to assure the protection and propagation of a balanced indigenous

population of shellfish, fish, and wildlife in and on the body of water into which the discharge is to be made, the Administrator (or, if appropriate, the State) may impose an effluent limitation . . . with respect to the thermal component of such discharge (taking into account the interaction of such thermal component with other pollutants), that will assure the protection and propagation of a balanced, indigenous population of shellfish, fish, and wildlife in and on the body of water.

Like new sources, dischargers of heat are entitled to a 10-year grace period from thermal effluent limitations stricter than modified limitations imposed after a 316(a) proceeding. This "balanced indigenous population" test for thermal variances has, if a pun may be excused, created more heat than light. Power companies, anxious to avoid the expense of cooling towers, have brought section 316(a) proceedings which are so complex that many remain unresolved after 15 years of administrative consideration. Once again, the water quality-based approach has failed because the effects of effluents on ecosystems are poorly understood.

Subsection 316(b) requires that in effluent limitations for thermal dischargers there be provisions "that the location, design, construction and capacity of cooling water intake structures reflect the best technology available for minimizing adverse environmental impact." This clause has also produced controversy, particularly with regard to EPA's approval of intake structures for the Seabrook nuclear plant in New Hampshire. Although the EPA regulates intake structures for nuclear power plants, the U.S. Supreme Court has held that radioactive discharges are under the jurisdiction of the Nuclear Regulatory Commission.

PROHIBITED POLLUTANTS

Few pollutants are absolutely interdicted. However, the CWA does forbid the issuance of permits for the discharge of radiological, chemical, or biological warfare agents and high-level radioactive wastes. Under section 307, the EPA has discretion to ban discharges of other pollutants that pose substantial dangers to public health. In general, when a herbicide or pesticide loses its registration under the Federal Insecticide, Fungicide, and Rodenticide Act, the EPA also bans its discharge into waterbodies. Discharges of polychlorinated biphenyls (PCBs) are prohibited under the federal Toxic Substances Control Act

At this point, the reader may find it helpful to examine Table 1, which illustrates the CWA's technology-based effluent limitations for industrial point sources by pollutant type, applicable effluent limitations, compliance dates, and available variances.

Table 1. Technology-Based Effluent Limitations

Pollutant Type	Eff. Limitation	Compliance Date	Variance
Conventionals	BCT	3 years from promulgation	FDF
Non-conventionals/ Nontoxics	BAT	same	FDF 301(c) and (g) variances based on economics and receiving water quality
Toxics	BAT	same	FDF 301(k) 2-year extension for adoption of innovative/ alternative technology
Heat	BAT	same	316(a) water quality-based variance 10-year grace period
New Source Discharges	BACT	on construction	none 10-year grace period

WATER QUALITY-BASED EFFLUENT LIMITATIONS

It may come as a surprise that the CWA perpetuated the water quality-based approach. After all, did not the failure of this strategy lead directly to the passage of the CWA with its emphasis on uniform national effluent limitations founded upon technological availability? Indeed, the original drafters of the 1972 act, the staff of the Subcommittee on Air and Water Pollution of the Senate Public Works Committee working under the supervision of

Senator Muskie, were committed to substituting the technology-based approach for the discredited water quality-based approach. As a result, the Senate bill did not provide for postenactment revision and approval of water quality standards. The House, on the other hand, was not willing to forego water quality standards. Congressman Blatnik, chairman of the House Public Works Committee and one of the sponsors of the House bill, had been one of the architects of the Water Quality Act of 1965, which codified the water quality standards approach. Also, there was feeling in the House that the novel technology-based effluent limitation approach might prove unsuccessful, and therefore water quality standards should be retained "just in case." Finally, members of the House envisioned "a dual approach; . . . whichever is the stronger shall apply." Thus, the House bill contained a section substantially identical to section 303 of the CWA, which the Senate was forced to accept—against its better judgment—by way of political compromise.

Section 303 requires that water quality standards be set for all waters, establishes procedures for the periodic review and revision of standards, and outlines a continuing water quality planning process for each state. Water quality standards must

> protect the public health or welfare, enhance the quality of water and serve the purposes of [the CWA]. Such standards shall be established taking into consideration their use and value for public water supplies, propagation of fish and wildlife, recreational purposes, and agricultural, industrial and other purposes, and also taking into consideration their use and value for navigation.

However, a state may not set a waterbody's designated use at less than fishable/swimmable without performing a use attainability analysis (see Chapter 26). The CWA does not specifically define fishable/swimmable, and a state is given latitude in applying this concept to segments within its borders. Having done so, a state must at least adopt specific numeric water quality standards for section 307(a) toxic pollutants as necessary to support designated uses where such pollutants are discharged or are present in the affected waters and could reasonably be expected to interfere with designated uses. These numeric standards may be based on section 304(a) criteria, drinking water standards, site-specific determinations, or other scientifically defensible methods. Where numeric criteria are unavailable, criteria based on bioassays must be adopted.

Each state is to reconsider its existing water quality standards at least every three years, revise them where appropriate, and submit revisions to the EPA for approval. The EPA must review the revised standards to determine whether they protect the public health or welfare, enhance the quality of water, and serve the purposes of the CWA. If the EPA determines that the revised standards meet the requirements of the act, the EPA approves the standards. The EPA must promulgate its own standards where it finds state revisions inadequate. The EPA is authorized not only to react to those revisions submitted by states, but to issue federal standards on its own initiative where appropriate.

Section 303 also establishes a procedure for the attainment of water quality standards. A state must first identify those waters within its boundaries for which technology-based effluent limitations are not stringent enough to attain water quality standards. These "water quality-limited stretches," as opposed to "effluent-limited stretches," are to be ranked in order of priority, taking into account designated uses and severity of pollution. The state must then set total maximum daily loads of pollutants (TMDLs) for water quality-limited stretches. Under section 303, "Such loads shall be established at a level necessary to implement the applicable water quality standards with seasonal variations and a margin of safety which takes into account any lack of knowledge concerning the relationship between effluent limitations and water quality." TMDLs must be submitted to the EPA administrator, who is authorized to set alternative loads if he finds a state's to be inadequate. Finally, TMDLs are to be converted to wasteload allocations and water quality-based effluent limitations as part of the state's continuing planning process.

Water quality-based effluent limitations must be imposed without regard to technological or economic considerations. Moreover, more stringent water quality-based limitations override all variances and grace periods from technology-based limitations except for section 316(a) variances, which the CWA expressly substitutes for water quality standards. Also, there is no lead time for meeting water quality-based effluent limitations; the CWA requires that they be achieved immediately. Economic disruption can only be avoided by revising water quality standards downward, a step that EPA regulations prohibit unless the economic consequences are "substantial and widespread."

Thus, the CWA combines the technology-based and water quality-based regulatory approaches. The problem is that this amalgam of effluent limitations contravenes one of the basic tenets of the CWA, discouraging "forum shopping" by industry through the development of uniform national effluent limitations based on available technology. The author has characterized this paradox as the "better than best" problem. Because most water quality-limited stretches will occur in heavily industrialized states, industry will be encouraged to site new plants in relatively less polluted or even high-quality areas. In order to attract industry, industrialized states will be compelled to revise their water quality standards. This will have a debilitating effect on the drive for clean water, because water quality standards serve a psychological, hortatory function as unmet goals. Revising standards connotes "writing off" a waterbody. A future siting scenario might find an industry playing off an industrialized state's promised revision against a less industrialized state's agreement to degrade high-quality waters. Existing industries might also use interstate disparities in effluent limitations to demand revisions or enforcement concessions. These industries would justifiably argue that they are being asked to bear a disproportionate pollution control burden not only as against their counterparts in other states but also as against nonpoint sources. Although nonpoint sources contribute as much to pollutant loadings as point sources, section 303 requires that wasteload allocations be set only for point source discharges, as if nonpoint pollution were insignificant. Moreover, the CWA does not adequately control pollution from nonpoint sources. In summary, retention of water quality-based effluent limitations promotes diversity in standard setting, and thus subverts the CWA.

The EPA has reacted to this dilemma by placing section 303 on the back burner. One rarely finds water quality-based effluent limitations in industrial discharge permits issued by the EPA or states. The EPA does not put pressure on states to develop TMDLs or set TMDLs for states that do not develop them, even though a court has ruled that the EPA bears a mandatory duty to set TMDLs for defaulting states. Moreover, the EPA has not provided adequate guidance to states on the politically controversial issue of converting TMDLs to water quality-based effluent limitations (i.e., making wasteload allocations). Vermont is one of the few states that has taken the wasteload allocation dilemma by the horns. In Vermont, assimilative capacity is allocated according to the following priorities: (1) existing discharges from POTWs treating existing sources; (2) existing discharges from other dischargers treating existing sources; (3) new discharges from POTWs originating from existing sources; (4) new discharges from other existing sources; (5) new discharges from POTWs treating new sources within their service areas; (6) new discharges from POTWs treating new sources outside their service areas; and (7) new discharges from other development. This is, in effect, prior appropriation of assimilative capacity.

Looking at the situation impartially, can anyone blame the EPA for its reluctance to enforce such an inefficient, inequitable, and politically inflammatory program? The author, among others,[11] has recommended that Congress repudiate section 303. The following strategies should be utilized to deal with water quality-limited stretches. For the long term, the EPA should take seriously its statutory responsibility to review BAT-based effluent limitations at five-year intervals. The nation should rely on progressively stricter but uniform technology-based effluent limitations to achieve cleaner water and ultimately zero discharge. At the same time, the EPA should revive and seek funding for its languishing water pollution control research program. This will assure that better technology becomes available. For the short term, the EPA should disinter section 302 of the CWA, which it has never used. Entitled "Water Quality Related Effluent Limitations," section 302 allows the EPA, after public hearing, to set stricter effluent limitations on heavily polluted waterbodies unless a discharger can show that he is entitled to a modification of these more stringent limitations. A discharger of nontoxic pollutants must show, to the satisfaction of the EPA and the state, "that (whether or not technology or other alternative control strategies are available) there is no reasonable relationship between the economic and social costs and the benefits to be obtained . . . from achieving such limitation." For dischargers of toxics, modifications are limited to one term of up to five years, and an applicant must demonstrate that the modification (1) will represent the maximum degree of control within his economic capability; and (2) will result in reasonable further progress toward compliance.

Section 302 would appear to be a feasible and fair means of resolving the "better than best" problem. Unfortunately, Congress reaffirmed section 303 in the 1987 CWA amendments. In a section entitled "Individual Control Strategies For Toxic Pollutants" (nicknamed "toxic hot spots"), the amendments provide for the development of water quality-based effluent limitations after the attainment of BAT-based effluent limitations.

States are required to undertake progressive programs of toxic load reduction where BAT is not sufficient to meet state water quality standards and support public health and water quality objectives of the CWA. Within two years of enactment, states must identify those waterbodies under their jurisdictions that will not meet state water quality standards because of toxic pollutants after the implementation of BAT-based categorical effluent limitations, new source performance standards, or pretreatment standards. For each identified segment, a state is to determine the specific point sources discharging toxic pollutants and the amount of each toxic pollutant discharged. The state submission is also to include an individual control strategy that the state determines will produce a reduction in identified toxic discharges, through the establishment of water quality-based effluent limitations derived from water quality standards containing numerical criteria. Where numerical criteria are unavailable, the state shall use bioassays. The state's proposed reduction in toxic discharges, in combination with other controls on point and nonpoint sources of pollution, must achieve the applicable water quality standard as soon as possible, but not later than three years after the effective date of the strategy.

Within 120 days after submittal, the EPA must approve or disapprove a state's control strategy. If a state does not submit its required control strategy within the two-year period, or if the EPA disapproves the strategy, then the EPA must, within one year, develop and implement a control strategy for the state.

No financial incentives are provided for states that implement their control strategies, and no penalties are included for states that do not, except for the hollow threat of EPA implementation. (The EPA lacks the resources to implement and enforce this program.) The 303 agony has merely been prolonged.

FEDERAL FLOORS AND STATE CEILINGS

The federal CWA requirements provide a "federal floor" that a state cannot undercut. For example, a state cannot impose an effluent limitation on a discharger that is less stringent than the federal technology-based limitation. On the other hand, a state can establish as high a ceiling as it can rationally sustain. Theoretically, a state could impose effluent limitations even more stringent than water quality-based limitations. And state-set effluent limitations are binding on the EPA. Realistically, states are reluctant to be overly strict with dischargers, as compared to their sister states. But, where the public is sympathetic, states frequently use their residuary powers to set more stringent limitations without running the 303 gauntlet.

NEW CONCEPTS IN INDUSTRIAL POINT SOURCE CONTROL

The EPA is considering several innovations in effluent limitations and water quality standards in order to reduce compliance costs. One, already promulgated for the iron and steel category, is the "bubble" concept. A bubble approach applies effluent limitations to plant discharges as a whole, instead of focusing on individual discharge points. Otherwise excessive discharges from one process can be set off against surplus effluent reductions from another process in order to reduce total treatment costs. In its effluent limitation regulations for the iron and steel category, the EPA has placed significant limitations on use of the bubble concept. If the facility elects to increase discharges at one pipe, compensatory reductions must be made at another pipe so as to achieve a net reduction of plantwide discharges of at least 10%. Thus far, the EPA has applied the bubble concept to only this one category, perhaps because its legality is questionable.

Similar to the bubble concept is that of "trading" surplus effluent reductions. Trading would either be "fixed" or "flexible," depending on whether or not a permit modification were necessary for each trade. Trading might be facilitated by central "banking" of surplus reductions. The EPA is studying not only potential trades between point source dischargers but also trades between point and nonpoint sources.[12] On the state level, Wisconsin regulations authorize trades between point source dischargers, provided the water quality standards are not violated.

Seasonal effluent limitations and water quality standards are being utilized by at least 40 states.[12] Most of these states divide the year into two seasons and use two sets of corresponding water quality standards and effluent limitations. A few states have adopted variable standards and limitations, keyed to more frequent changes in receiving water conditions. Site-specific standards and limitations are becoming increasingly popular. For example, Pennsylvania has six different dissolved oxygen levels that may be applied to any one of seven aquatic protection uses, based on a cold water-warm water division. Montana has adopted water quality standards for four different kinds of trout streams.

Two comments deserve to be made concerning these innovative approaches. First, most of them require use attainability analyses, placing an extraordinary administrative burden on the states and EPA. Second, they are all ways of dealing with section 303, facilitating lower standards and limitations at various places and times on water quality-limited stretches. As such, they presuppose the continued validity of the water quality-based approach after the attainment of technology-based effluent limitations. The EPA's acceptance of these strategies and Congress' enactment of the "toxic hot spots" amendment leave no doubt that the water quality-based approach is alive and well—is in fact making a comeback.

Chapter 28

THE CLEAN WATER ACT: MUNICIPAL EFFLUENT LIMITATIONS

Publicly owned treatment works are also subject to technology-based effluent limitations and, on water quality-limited stretches, to water quality-based effluent limitations. The technology-based limitations for POTWs are based on "secondary treatment," and the attainment date is July 1, 1988, postponed by Congress from 1977 and 1983. POTWs must comply regardless of whether their owners have received federal construction grants.

"Primary treatment" involves the use of mechanical means, such as screens, grit channels, and settling tanks, to remove large materials and suspended solids from the largely liquid waste stream. The process removes about 30% of biochemical oxygen demand from sewage. "Secondary treatment" entails additional settling and placing the sewage in contact with microorganisms, by way of trickling filters, activated sludge, or lagoons, to break down the organic matter in the sewage stream. An additional 50–60% of BOD, most suspended solids, and some toxic materials, that enter the sewerage system mainly from indirect dischargers and street runoff, are removed at this stage. The liquid effluent from secondary treatment is generally chlorinated and discharged. Sludge residues of primary and secondary treatment are either incinerated, spread on land, buried in landfills, dumped in the ocean, or composted for use as a soil conditioner. Advanced waste treatment (AWT), sometimes referred to as "tertiary treatment," refers to processes that remove additional pollutants from wastewater beyond those eliminated by primary and secondary treatment. The AWT processes may remove nutrients or higher percentages of BOD and suspended solids.

Section 301(h), added to the CWA in 1977, authorizes the Environmental Protection Agency, with the consent of a state, to modify the act's secondary treatment requirement for a POTW that discharges into marine waters if the applicant can show that

> (1) there is a water quality standard specific to the pollutants for which the modification is requested, and the modified discharge would not violate the standard; (2) the modified requirement will not interfere with the attainment of fishable-swimmable waters, or with the existing ecosystem; (3) a practicable monitoring system has been established to measure the impact of the modified discharge on aquatic biota; (4) the modified discharge will not place added burdens on other dischargers; (5) controls have been placed on indirect dischargers to the system; (6) nonindustrial toxics are, to the extent practicable, being denied entrance to the system; and (7) the modified effluent limitation will not be exceeded.

Section 301(h) defines "marine waters" as ocean and coastal waters as well as "saline estuarine waters where there is a strong tidal movement and other hydrological and geological characteristics" that the EPA decides are necessary to protect water quality. Modifications are unavailable to ocean dischargers of sewage sludge.

In the 1987 CWA amendments, Congress placed further restrictions on the granting of 301(h) modifications: (a) in the case of a POTW serving 50,000 people or more, where there is no applicable pretreatment standard, pretreatment of toxic pollutants must be adequate to achieve the equivalent of secondary treatment for the overall system; where a pretreatment standard exists, it must be met and enforced; (b) primary treatment must be achieved "after initial mixing"; (c) water providing dilution must not contain significant amounts of previously discharged effluent from the POTW; (d) estuarine receiving waters must not be impaired environmentally at the time of application; and (e) no variances may be granted for discharges into the New York Bight.

As of mid-1986, the EPA had received 208 applications for 301(h) modifications. Of these, approximately 50 applications were approved, all from dischargers on the West Coast where dilution is greater than on the East Coast. All applications for discharges into estuarine waters were denied, and all dischargers granted modifications were providing at least primary treatment.[13]

Controlling stormwater discharges has been a recurring problem under the CWA. Stormwater outlets and overflow points of combined sewers are point sources, and courts have ordered the EPA to control them, but the magnitude of the task has paralyzed the EPA.

In the 1987 amendments, Congress enacted new timetables and standards for controlling discharges from separate storm sewers. Prior to 1992, no permit may be required for a municipal stormwater discharge except where: (1) a permit has already been issued; (2) the discharge is associated with an industrial activity; (3) the discharge is from a municipal separate storm sewer system serving a population of 250,000 or more; (4) the discharge is from a municipal separate storm sewer system serving a population of 100,000 or more but less than 250,000; or (5) the discharge contributes to a violation of water quality standards or is a significant contributor of pollutants.

Discharges from category number 5 must be addressed immediately. For discharges covered by 2 and 3, applications for permits must be submitted by 1990, the permits must be issued or denied by 1991, and compliance must be achieved by 1994. For discharges covered by 4, the applicable dates are 1992, 1993, and 1996. After October 1, 1992, the permit requirements of the CWA are restored for municipal separate storm sewer systems serving a population of fewer than 100,000 persons.

General permits may be issued for discharges from municipal storm sewers. Either general or individual permits must include a requirement to effectively prohibit nonstormwater discharges into the storm sewers. Stormwater permits "shall require controls to reduce the discharge of pollutants to the maximum extent practicable, including management practices, control techniques, and system design and engineering methods, and such other provisions as the Administrator or the State determines appropriate for the control of such pollutants."

Combined sewer overflow point discharges were not considered in the 1987 amendments. The applicable control standard is best practicable technology currently available, but there have been few efforts at control by the EPA or the states.

Chapter 29

THE CLEAN WATER ACT: FEDERAL SUBSIDIES FOR POTW CONSTRUCTION

This area of the law is in a transition phase. The federal construction grants program, established in 1956 and authorized at $5 billion per year at its height, is gradually being replaced by a program of federal capitalization grants for State Revolving Loan Funds (SRFs). These programs will be administered concurrently at least until 1990; consequently it is necessary, at this time, to be familiar with the legal contents of both programs.

CONSTRUCTION GRANTS PROGRAM

Eligibility and the Federal Share

The construction grants program has been a dynamic one, responding to changes in the political climate. This is especially clear with regard to eligibility for federal funding and the amount of required state and local cost-share.

Various kinds of POTW construction-related projects have been eligible for funding under the act: secondary treatment; AWT where necessary; new interceptor sewers; infiltration-inflow correction; new collector sewers; replacement and rehabilitation of sewers; correction of combined sewer overflow; facilities for storage and recycling of wastes; and acquisition of lands needed for the facilities themselves or for sludge or effluent disposal. In 1977, recognizing that in some areas community managed septic systems are an environmentally sound and cost-effective alternative to sewage treatment plants, Congress authorized grants for privately owned systems in existing communities where a public agency agrees to control the design, installation, and operation of septic systems constructed with grant funds.

Although these categories have been legally grant-eligible, they have not been funded equally. For example, the Environmental Protection Agency traditionally funded few projects to correct combined sewer overflows and, in recent years, has restricted advanced waste treatment funding to projects on water quality-limited stretches that will produce significant water quality benefits. In 1977, Congress declared that each state must use at least 25% of its alloted construction grant funds for major sewer rehabilitation, new collector sewers, new interceptors, and correction of combined sewer overflows if projects in these categories appear on a state's priority list. However, in 1981 Congress again changed the eligibility rules with regard to grants approved after October 1, 1984. After that date only secondary treatment, AWT, new interceptor sewers, and infiltration-inflow correction have been directly eligible. But a state governor may use up to 20% of a state's allotment for other kinds of projects. The 1981 amendments also instituted a separate grants program, authorized at $200 million annually, to correct combined sewer overflows from urban areas into marine bays and estuaries. In 1987 Congress enacted a preallotment set-aside of 1% of all sums appropriated for fiscal 1988, and 1½% per annum for fiscal 1989 and 1990, for the same purpose.

One of the controversies manifested by these shifts in funding priorities has been over the extent to which the federal government should be subsidizing community growth. Section 204 of the CWA states that projects funded by construction grants should include "sufficient reserve capacity." Since early in the administration of the construction grants program, environmentalists have charged that funding reserve capacity encourages sewerage officials, in the name of speculative growth, to overbuild treatment plants and construct far-flung sewer systems which cause energy-intensive, environmentally destructive, and infrastructure-demanding "leapfrog development." Before 1981, all environmentalist attempts to have Congress limit funding for reserve capacity were defeated by development interests. However, in 1981 a combination of environmentalist pressure and the Reagan administration's fiscal policies persuaded Congress to place restraints on federal funding of reserve capacity. All segments and phases of treatment plants and interceptor lines funded after 1981 and prior to October 1, 1984, were funded based on the 20-year reserve capacity applicable to earlier projects. After 1984, no grant can be made to provide reserve capacity in excess of the needs existing on the date of the award or those existing on October 1, 1990, whichever comes first. In short, states and municipalities must now finance their own POTW facilities for anticipated growth.

Restrictions on POTW-induced growth can also be found in section 316 of the Clean Air Act. Before awarding a construction grant, the EPA must be satisfied that the direct and indirect air pollution emissions likely to result from the construction of a POTW will conform to the requirements of a state implementation plan. Moreover, grants may be withheld in non-attainment areas.

Decreasing the federal share has also caused local governments to be more cautious in planning for growth. The 1972 CWA increased the federal share of POTW financing from 55% to 75%. After October 1, 1984, the 55% ceiling was reinstated, except for the later stages of projects funded before that date. Because a number of municipalities found it increasingly difficult to raise their local shares of POTW construction costs, the CWA was amended to authorize federal guarantees of municipal financing where necessary.

The CWA construction grants program has been the major source of federal funds for POTW construction, but two other federal programs have also assisted municipalities to construct facilities or raise their local shares. The Farmer's Home Administration makes grants and loans to community and nonprofit groups in rural areas with populations of less than 10,000 to construct centralized and on-site water supply and waste treatment systems. The Department of Housing and Urban Development makes community development block grants available to municipalities, which can use the proceeds for construction or cost-sharing.

Innovative and Alternative Waste Treatment

Environmentalists have criticized traditional secondary treatment for wastefulness because, for one thing, secondary plants discharge water derived from groundwater aquifers to rivers flowing into an ocean or the Great Lakes. Instead, environmentalists have recommended alternatives such as spraying effluent, after primary treatment and chlorination, on land and allowing the "living filter" to remove impurities while groundwater supplies are replenished.

In 1977 Congress passed a number of amendments to stimulate innovative and alternative (I/A) waste treatment. No construction grant can be made unless the applicant demonstrates that

> innovative and alternative wastewater treatment processes and techniques which provide for the reclaiming and reuse of wastewater, otherwise eliminate the discharge of pollutants, and utilize recycling techniques, land treatment,

new or improved methods of waste treatment management . . . and the confined disposal of pollutants . . . have been fully studied by the applicant taking
into account . . . the more efficient use of energy resources.

I/A projects that conform to EPA guidelines may be funded even
though they cost up to 15% more than the most cost-effective alternative.
Moreover, the federal grant share for I/A projects is 75% instead of 55% for
traditional treatment works, and extra grant monies will be available for
modification or replacement of the funded facilities if they do not meet design
specifications, except where negligence has occurred. Furthermore, a minimum of 4% and a maximum of 7.5% of each state's allotment must be set
aside for I/A bonuses. There is also a 7.5% maximum "set aside" for funding alternatives to conventional treatment works in communities of fewer
than 3500 people located in states with rural populations of 25% or more.

According to the United States General Accounting Office, the I/A program has had limited success.[14] It does not provide sufficient incentives for
consulting engineers and states to take the risk or incur the additional cost
of developing innovative projects. The GAO recommended more centralized
review of I/A applications and a federal demonstration program for potential innovative technologies. The state of Florida has enacted a statute
insulating persons who use treated wastewater for spray irrigation from
liability, unless they have been negligent.

Private industrial firms are also rewarded for using innovative technology. Any company proposing to install innovative production processes
or techniques that would achieve significantly greater effluent reductions
than called for by the applicable effluent limitations, or that would achieve
categorical effluent limitations at a significantly lower cost than existing systems, may be granted a two-year extension of its date for compliance with
BAT-based effluent limitations if the innovative systems possess the potential for industry-wide application.

Innovative and alternative waste treatment processes involve replen-
ishing natural systems or recycling and reusing water by POTWs or industry.
Either way, discharges to waterbodies will be diminished and downstream
or lake diversion rightholders may lose their supplies. As mentioned in Parts
I and III, downstream appropriators have no rights to the continuation of
a POTW discharge. However, the rule may not be the same for industrial
discharges, which may be treated as irrigation return flows. The issue is further complicated by the presence in the CWA of section 101(g): "It is the
policy of Congress that the authority of each State to allocate quantities of
water within its jurisdiction shall not be superseded, abrogated, or otherwise impaired by this Act." Does this mean that effluent limitations cannot
interfere with prior diversion rights? Is it merely an unenforceable affirma-
tion of state primacy as to water allocation? Subsection 101(g)'s legislative
history is so sparse that courts may find these issues difficult to resolve.

Thus far, courts have not interpreted section 101(g) as possessing independent substantive effect.

The Funding Process and Its Administration

There has been a trend since 1972 toward state administration of the construction grants program. Prior to 1977, each state developed a "priority list" of needed projects and submitted it to the EPA. In turn, the EPA made the final decisions about how the state's allotment was to be spent among the priority projects and supervised the planning and construction of POTWs with the assistance of state officials. But in 1977 Congress declared a policy "that the States manage the construction grants program," and authorized delegation of program administration to states. The EPA was empowered to reserve up to 2% of a state's allotment for making program administration grants to a state accepting delegation. In 1981 the "administration set aside" was increased to 4%. State priority lists are now binding on the EPA, unless "the administrator, after a public hearing, determines that a specific project will not result in compliance with the enforceable requirements" of the CWA; in other words, unless the project as designed will not meet its effluent limitations. Where a state has been delegated management authority, the EPA must approve or disapprove a state grant certification within 45 days or the grant application is deemed approved. The details of delegation, including how and when the EPA will exercise its powers of review and supervision, are included in a "delegation agreement" between the EPA and the state. From time to time it is recommended that the EPA be eliminated from the construction grants program completely, and that allocations be made directly to states.

The construction grants funding process was significantly changed in 1981. Before that, except for very small projects, POTWs were funded in advance by the EPA in three steps: (1) facilities plans; (2) engineering studies, architectural designs, drawings, and specifications; and (3) construction. Each of these three steps was considered as a separate project and had to be funded individually by the EPA. In order to encourage local responsibility for POTWs and discourage overbuilding, Congress amended the CWA in 1981 to require local financing of steps 1 and 2, with federal reimbursement for the reasonable amount of these costs if and when the EPA approves a step 3 grant. For small communities that need assistance in financing steps 1 and 2, states are required to make loans of up to 10% of their allotted funds, with the amount of each loan subsequently deductible from construction grants if any are received. The 1987 amendments initiated an expedited procedure for constructing small, low-technology projects costing less than $8 million, called "Design/Build Projects."

Facilities planning has assumed an importance unforeseen in 1972. The drafters of the CWA intended that siting and sizing of POTWs be consistent with comprehensive "water quality management plans" required by section 208 of the act. However, because of delays in the 208 areawide planning process, facilities plans have become major instruments for evaluating the considerable environmental impacts of regional treatment facilities.

The term "facilities planning" does not appear in the act. It is derived from section 201(c), which stipulates that "to the extent practicable, waste-treatment technology shall be on an areawide basis and provide control or treatment of all point and nonpoint sources of pollution." Municipal facilities must also be cost-effective, and the EPA is required to encourage the construction of revenue-producing POTWs. As a result, the facilities plan has become an often uneasy combination of cost-effectiveness and environmental impact evaluation.

A facilities plan must be prepared for the entire service area under consideration. The components of a facilities plan are:

(1) A description of the complete waste-treatment system, as well as a description of the specific collection and treatment works for which construction plans and specifications are to be prepared during step 2.

(2) An extensive analysis of any infiltration/inflow problems that might be present in the existing system and a plan for rehabilitating the leaking sewers.

(3) An analysis of the cost-effectiveness both of the waste-treatment system of which the proposed project is a part and of alternative systems, including innovative systems. This analysis must describe the relationship of the capacity of the system to the needs to be served, and the degree to which effluent quality could be improved by upgrading existing facilities rather than by constructing new ones. Alternative methods of effluent and sludge disposal must also be considered.

(4) An environmental impact assessment of all foreseeable impacts of the project and reasonable alternative projects.

Facilities plans are prepared by consulting firms retained by the applicant. Fees of these consulting firms are based on a percentage of the project cost. Environmentalists have cited these percentage fee arrangements as one reason for enacting statutory limitations on POTW size and reserve capacity.

The facilities plan must be approved by the construction grants management agency, either the EPA or a state, before a project can move forward. Public participation is mandated during facilities planning, and comments elicited at public hearings have frequently drawn the attention of state and federal officials to defects in the plan. A federal construction grant is one of the two EPA water quality-related actions (the other is EPA issuance of a new source discharge permit) that are covered by the National Environ-mental Policy Act. The facilities plan is generally the core of the EPA's environmental assessment and Finding of No Significant Impact, or in some cases of a full-fledged environmental impact statement.

As section 208 plans have been completed, facilities planning has been amalgamated, or incorporated by reference, into 208 planning (see Chapter 36). After a 208 plan has been approved, all construction grants must be made to the designated 208 agency and only for works in conformity with the plan. Under the 1987 amendments, if a 208 plan is incomplete, a con-struction grant can only be made "if a plan is being developed for such area and reasonable progress is being made toward its implementation and the proposed treatment works will be included in such plan." The POTW must also be in conformity with an existing or uncompleted state plan.

User Charges, Industrial Users, and Performance Safeguards

Two features of the CWA as originally enacted, "user charges" and "industrial cost recovery," provoked controversy almost immediately. In order to obtain a federal construction grant, a municipality must adopt a system of user charges ensuring that all recipients of waste-treatment serv-ices pay their proportionate share of operating and maintenance expenses. Under the 1972 act, a user charge system had to be based on volume and concentration of wastes entering the POTW. However, many municipali-ties were unable to set user charges based on influent volume and concen-tration because they lacked metering capability, especially for residential and commercial users. In 1977 the CWA was amended to allow user charges for residential and commercial establishments to be based on property value. In order to qualify, an ad valorem levy must have been in effect before 1977, must produce revenues dedicated to operation and maintenance of the POTW, and must allocate the burden proportionately both within and among classes of users. Quantity discounts for large-volume users are unaccept-able, but preferential rates for low-income users may be included. Metering, however, is still required for industrial users whose flows exceed 25,000 gallons per day. Indirect dischargers of toxic materials must pay for increased treatment and sludge-handling costs.

Originally, an industrial user of a POTW (indirect discharger) was required to repay a portion of the federal grant corresponding to its percen-tage use of the system's total capacity. "Industrial cost recovery" was intended to provide funds for reconstruction and repair, and to encourage an industry to choose the most cost-effective solution to its waste treatment problems, whether it was indirect or direct discharge. After prolonged and concerted opposition from municipalities and sewerage authorities, indus-trial cost recovery was first suspended, then limited to large indirect dischargers, and finally repealed entirely. Thus, as the CWA now stands, there is a major federal subsidy for industries that "plug in" to POTWs, because they pay only user charges and therefore save on construction costs. Furthermore, direct dischargers that contract to plug in have been granted extensions of time to meet applicable effluent limitations. These major incentives for indirect dischargers emphasize the need for an effective industrial pretreatment program.

Industrial waste overloads are one reason why many POTWs built under the construction grants program are not performing up to expectations.[15] Other problems have been design deficiencies, equipment deficiencies, infiltration/inflow problems, and deficiencies in operation and maintenance. In its 1981 amendments to the CWA, Congress took the first steps in dealing with this embarrassing situation. Before a construction grant is made, the EPA "shall determine that the facilities plan . . . constitutes the most economical and cost-effective combination of treatment works over the life of the project to meet the requirements of [the CWA]." Water conservation methods are to be considered in determining cost-effectiveness. A "value engineering review" must also accompany each application.

The CWA's most powerful safeguard against POTW inadequacy involves continuing responsibilities of the prime engineering consultant. Grant funds are provided to retain the prime consultant on the site for one year after startup to supervise operations and train personnel. At the end of that time, the consultant must assist the POTW's owner or operator in certifying whether or not the POTW meets design specifications as well as applicable effluent limitations. If the plant does not comply, remedial measures must be taken "in a timely manner" at the grantee's expense. However, compliance after one year does not ensure that compliance will continue. The General Accounting Office has found that user charges are often insufficient to pay for replacement.[16] In addition, the shortage of trained plant operators has been a dilemma. The EPA could revise its user charge regulations to improve POTW self-sufficiency and integrity. As for operator shortages, the EPA could request and allocate more funds for operator training and scholarship grants to universities under sections 109 and 111 of the CWA.

Thus far, Congress has done nothing about problems that are caused by faulty construction, as opposed to faulty design, of POTWs. Where the EPA is the management agency it has utilized the Army Corps of Engineers to prevent defective construction, such as improperly laid sewer pipes, which are susceptible to infiltration and inflow. But, states that have accepted delegation may not possess the resources to competently supervise plant construction without strong program funding and technical support from the EPA. Also, state laws may not provide effective sanctions to deter contractors from taking shortcuts.

Privatization of Sewage Treatment

A novel concept that has been bruited recently is the privatization of sewage treatment; that is, allowing the private sector to take over all or part of what has long been a public function. Given the restrictions imposed by municipal bond law, it is doubtful that POTW ownership could be transferred to private firms. But, where POTWs do not already exist, companies might be given public utility status to provide sewerage services. More likely, consulting firms will be retained to manage, operate, and maintain regional POTWs. However, recent statutory restrictions on using tax exempt bond proceeds for private activities will undoubtedly discourage these types of arrangements.

GRANTS FOR STATE REVOLVING FUNDS

Beginning in 1989, federal funds have been authorized for capitalization grants to states to establish State Revolving Funds (SRFs). Authorized amounts are $1.2 billion per annum for fiscal 1989 and 1990, $2.4 billion for fiscal 1991, $1.8 billion for fiscal 1992, $1.2 billion for fiscal 1993, and $600,000 for fiscal 1994. Until 1990, a state's governor may elect to have a certain percentage of his state's construction grant allotment deposited in its SRF.

Any state desiring to establish an SRF must enter into a capitalization grant agreement with the EPA. The state must agree to provide a 20% cash match of federal funds and make binding commitments for assistance in an amount equal to 120% of the amount of the grant payment within one year after receipt.

The state must use its SRF first to assure that POTWs maintain progress toward compliance with the enforceable deadlines, goals, and requirements of the CWA, including the municipal compliance deadline of July 1, 1988; or, that they are on enforceable compliance schedules after that date. Once having met this requirement, the SRF may be used for other treatment works, nonpoint source management programs, and estuary protection.

Any POTW constructed wholly or partly with funds made available by capitalization grants (i.e., not exclusively with state matching funds, repayment monies, or other funds) must meet the following requirements: (a) the project must be for secondary or more stringent treatment or any cost-effective alternative thereto, new interceptors, infiltration-inflow correction, or correction of combined sewer overflows where they are a major priority; nevertheless, a governor may reserve 20% of a state's capitalization grant for other water pollution control projects; (b) innovative and alternative treatment technologies must be studied; (c) the related collection system must not be subject to excessive infiltration; (d) recreation and open space opportunities at the proposed facility must be studied; (e) the project must be included in, and consistent with, plans developed under sections 208 and 303(e); (f) communities constructing projects must develop user charge systems and possess the legal, institutional, managerial, and financial capability to construct, operate, and maintain the POTW; (g) one year after construction, the owner or operator must certify that the facility meets its design specifications and discharge permit effluent limitations; (h) projects must be cost-effective, and projects over $10 million must include a value engineering report.

Where POTWs are concerned, states may use their SRFs to (a) make low interest or interest-free loans, for terms not to exceed 20 years, with all payments of principal and interest credited to the SRF (recipients must establish dedicated sources of revenue for repayment); (b) buy or refinance debt obligations of construction agencies at below market rates; (c) guarantee, or purchase insurance for, local obligations in order to improve market access or reduce interest rates, (d) act as security for state general obligation bonds, if the bond sale proceeds will be deposited in the SRF; (e) provide loan guarantees for substate revolving funds; (f) earn interest; and (g) pay for SRF administration, up to 4% of grant awards.

State Revolving Fund financial assistance may only be made to a project on the state priority list, although rank priorities on the list need not be followed. Assistance may be provided with respect to the nonfederal share of a construction grant-funded project, but only where such assistance is necessary to allow the project to proceed.

The EPA must review each SRF at least annually. Where a state is misusing its SRF, and does not take corrective action, the EPA may withhold additional payments and eventually reallot the state's initial allotment.

EVALUATION OF MUNICIPAL POINT SOURCE CONTROL PROGRAM

The municipal point source control program under the Clean Water Act has been far less successful than its industrial counterpart. In early 1986, approximately 50% of all POTWs had not achieved compliance with the act's secondary treatment requirement, and it seems clear that many of these facilities will not meet the July 1, 1988 compliance date. Many cogent reasons have been given for this relative failure: insufficient federal construction grants funding; inability of municipalities to raise construction funds; repeal of industrial cost recovery inhibiting repair and reconstruction; inadequate user charges causing poor operation and maintenance; poor plant design and construction; lack of trained operators; and poor enforcement by the EPA. But at least as important has been the nearly complete failure of the pretreatment program.

Chapter 30

THE CLEAN WATER ACT: INDUSTRIAL PRETREATMENT

"Industrial pretreatment" means the treatment of industrial wastes before they are discharged into publicly owned treatment works. Pretreatment is necessary because certain indirect discharges are "incompatible" with a POTW. This means that they interfere with the POTW's operation, contaminate the sewage sludge, or pass untreated through the POTW in quantities so large that the POTW is in violation of its effluent limitations. Estimates vary regarding the number of industries discharging incompatible pollutants into POTWs, but there are probably around 14,000. Although only a small number of POTWs, perhaps 1,700 out of almost 17,000 POTWs nationwide, have problems with incompatible pollutants, these POTWs generally discharge into highly polluted waterways.

Controlling indirect discharges of incompatible pollutants has been one of the CWA's major failures. According to the General Accounting Office, "almost from its inception, the pretreatment program has been surrounded by controversy and has generated considerable uncertainty and confusion."[17] The pretreatment program's stagnation is a function of two myths which are common among regulators and regulated alike. The first myth is that municipalities and sewerage authorities will regulate incompatibles in order to protect the treatment process. This is undoubtedly true in the case of incompatibles that interrupt the biological process, but does not apply to pollutants that pass through into the effluent or end up as part of the POTW's sludge. The second myth is that POTW owners will impose pretreatment in order to meet POTW effluent limitations and sludge-quality regulations.

However, until approximately 1985 federal and state enforcement against POTWs and indirect dischargers was virtually nonexistent. And the EPA's sludge management regulations have not yet been promulgated. All other things being equal, POTWs will not enforce against their own customers, who will threaten to move elsewhere, recycle, or switch to direct discharge if pretreatment costs increase substantially.

Discharging toxic pollutants into sewer lines is attractive to industry for reasons other than the overall weakness of the pretreatment program. First, there is a "domestic sewage exclusion" in the Federal Resource Conservation and Recovery Act that exempts anything discharged into a sewer from RCRA's reporting, permitting, and treatment requirements (i.e., it is not treated as a "hazardous waste"). Second, the CWA does not require discharge permits for indirect dischargers; some states do, but most do not. Finally, since the repeal of industrial cost recovery, indirect discharge is a distinct bargain.

NATIONAL PRETREATMENT STANDARDS

There are two types of pretreatment standards: "prohibited discharge standards" and "categorical pretreatment standards." Prohibited discharge standards require that pollutants introduced into a POTW not inhibit or interfere with the POTW operations or performance. For example, it is illegal to introduce into a POTW explosives, corrosives, slug discharges, and solids and heat in excessive amounts. These standards apply to all nondomestic (commercial and industrial) users, whether or not they are subject to other national, state, or local pretreatment requirements. Categorical pretreatment standards set out national discharge limits for priority pollutants (toxics) based on BAT, as required by the Consent Decree and the CWA. Separate regulations containing pretreatment standards have been developed and promulgated for each of 42 specific industrial categories. These national pretreatment standards contain exemptions for small indirect dischargers in each category. Pretreatment standards for combined waste streams are set by applying a specified formula to mixed waste streams where at least one of them would otherwise be subject to pretreatment standards. The CWA requires compliance with categorical pretreatment standards within three years from their effective date.

Categorical pretreatment standards are based on the treatment technology available to industry. They do not recognize that traditional secondary treatment is effective in removing some priority pollutants, especially organic chemicals (some of which, unfortunately, become air pollutants). In order to obviate redundant treatment, the CWA was amended in 1977 to authorize "removal allowances" for indirect dischargers subject to categorical pretreatment standards. Removal allowances are based on the proven ability of a POTW to treat toxic pollutants; dilution in the plant or sewerage system cannot be taken into consideration. Thus, if the POTW consistently removes a pollutant, the industry-wide indirect discharge limits applicable to dischargers to that system may be raised for the removed pollutants. The burden of developing and granting removal allowances is on the POTW. Environmental Protection Agency regulations stipulate that the POTW must demonstrate "consistent removal of each pollutant" and provide continued monitoring of its removal efficiency in order to qualify for a removal allowance. The CWA conditions a removal allowance on (1) the POTW removing all or part of the pollutant; (2) the POTW not violating the effluent limitation for the incompatible pollutant that would be imposed on the indirect discharger if it were a direct discharger; and (3) the POTW's sludge not violating CWA standards. The removal allowance must be discontinued if the POTW ceases to meet the requirements and fails to take appropriate corrective actions.

The General Accounting Office has sharply criticized removal allowances as being administratively unworkable for POTWs and inequitable to industry "because companies in the same industry could be required to meet different standards for the same pollutants—depending on the POTW's pollutant removal efficiencies and the POTW's willingness to apply for the allowance."[17]

The EPA has promulgated four sets of removal credit regulations since 1973. Each set of regulations has been more lenient toward POTWs and industry than its predecessor, so as to encourage POTWs to apply for removal credits. Finally, the Natural Resources Defense Council (NRDC) challenged the EPA's 1984 regulations as violating the CWA in four respects. The Third Circuit, in *NRDC* v. *EPA*,[18] agreed with all four of NRDC's claims:

Consistent Removal Rate

The EPA's definition of "consistent removal rate" has been progressively weakened since 1973: from removal that occurs 95% of the time, to 75% of the time, to 50% of the time (average amount of 12 monthly samplings). In the first place, said the Court, "(w)e find it difficult to fathom how a level of removal that is met one half of the time and exceeded one half of the time, and that contains no limit on the permissible amount of variability, can be termed consistent.'" Secondly, the EPA's conclusion that an averaging approach would not cause an unacceptable amount of water pollution was based on a false assumption: that a POTW will rarely perform substantially below its average. On the contrary, the evidence showed marked variability in POTW removal performance. "These data clearly reveal that a POTW may remove virtually all of a particular pollutant on one day while removing little or none of that pollutant on day two."

Combined Sewer Overflow Adjustment

Prior to 1984, the EPA's removal credit regulations reduced an available removal credit by a percentage equal to the frequency of combined sewer overflows experienced by the POTW, on the ground that overflows cannot be considered consistent removal. This adjustment was dropped in 1984. The EPA reasoned that combined sewers will overflow an average of 7.3% of the time, and that a 7.3% adjustment to a removal credit would be insignificant. Once again the Court took the EPA to task for using averages improperly. Philadelphia, for example, has overflows 13% of the time. Moreover, a 7.3% reduction in pollutants discharged may have a significant effect on water quality.

Withdrawal of Removal Credits

The 1984 rule changed the test for determining when a removal credit has to be modified or withdrawn. Instead of acting whenever the POTW no longer fulfills the conditions justifying a credit, the EPA would, under its 1984 rule, withdraw a credit before the expiration of a POTW's five-year permit only if the POTW's removal rate drops "consistently and substantially" below the rate claimed in the permit application. The Court held this test to be an obvious violation of the "consistent removal" mandate, in effect standing consistent removal on its head.

Consistency with Sludge Regulations

Removal credits may not be issued where the POTW's sludge would violate the EPA's sludge utilization and disposal regulations. But the EPA had not then promulgated these rules. The NRDC argued that this failure precluded the authorization of removal credits. The Court agreed, brushing aside the EPA's contention that regulations promulgated under other statutes were the functional equivalent of sludge utilization and disposal rules. "Certainly, the existing regulations do nothing to advance the congressional goal of making sludge into a productive asset rather than a toxic liability."

The Third Circuit's NRDC v. Train decision threw the pretreatment program into even further disarray. Inexplicably, Congress only addressed the sewage sludge regulations aspect of this decision in its 1987 amendments. This part of the decision was stayed until August 31, 1987—when the EPA was required to promulgate sludge management regulations—for those POTWs that had already applied for or received approval to issue removal credits. For other POTWs, no removal credits could be issued until the regulations were promulgated. After the 1987 amendments, the U.S. Supreme Court refused to review the Third Circuit's decision. Thus, pretreatment is still "on hold" because the Third Circuit's other objections to the EPA's removal credit regulations remain viable. In addition, the EPA's sludge management regulations are not expected until 1990.

LOCAL PRETREATMENT PROGRAMS

The CWA's pretreatment strategy envisions a parallel effort on the part of federal, state, and local governments to implement pretreatment requirements. Local pretreatment programs, approved by states, are the focus of this effort. States, through EPA-approved state pretreatment programs, are to oversee the operation of the local programs, provide backup enforcement capability, and, where a local program has not been developed or is not being enforced, assume primary responsibility for applying pretreatment standards to industrial users. The EPA is to supervise state pretreatment programs, provide backup enforcement to state and local governments, and, in the absence of a viable local or state pretreatment program, function as the primary enforcement agency. POTWs with a total design flow of 5 million gallons per day or greater and receiving incompatible pollutants must have had a pretreatment program approved by July 1, 1983. No POTW may grant a removal allowance unless it possesses an approved pretreatment program. States must also have had their pretreatment programs approved by July 1, 1983. Both program and construction grant funds may be used by states to develop and implement pretreatment programs. However, the GAO has concluded that "the ability of POTWs, States, and EPA to meet the substantial resource commitment that the Federal pretreatment program will require is highly questionable."[17] Perhaps more to the point, there is the "fox in the henhouse" problem of asking local governments to regulate their customers.

Chapter 31

THE CLEAN WATER ACT: NATIONAL POLLUTANT DISCHARGE ELIMINATION SYSTEM

Effluent limitations are not in themselves enforceable against industrial or municipal point source dischargers. In order to be enforceable, they must be translated into permit conditions. Under the Clean Water Act, it is illegal to discharge without a permit. A discharge permit is not a "license to pollute." It is either a temporary privilege to use a waterbody for waste disposal until improved pollution control technology is developed or, where a water quality-based variance is available, a warning that no discharge will be tolerated if it disrupts the integrity of the waterbody. This strong presumption against using waterbodies for waste disposal is illustrated by the title of the CWA's permit system, the "National Pollutant Discharge Elimination System" (NPDES). The NPDES program is administered either by an EPA regional office or a state that has received EPA permission to issue permits.

Every operator of an existing discharger or potential operator of a new source must apply for an NPDES permit. The application must contain accurate discharge information, including the results of tests specified in the regulations. Once a permit application is complete, the permit authority must decide whether to deny the permit or to prepare a draft permit. A draft

permit must be accompanied by a fact sheet and a public notice that specifies how comments can be made and how a public hearing can be obtained
if there is sufficient public interest. Once a decision is made to issue a permit,
the agency must issue a final written decision that responds to comments.
Only new source permits issued by the EPA are subject to the National
Environmental Policy Act. Interested persons who have filed comments may
take an administrative appeal if they are dissatisfied with the result. Unlike
the public hearing, this evidentiary hearing is what lawyers refer to as "on
the record;" that is, limited cross-examination of witnesses is allowed. Further appeals may be take to the administrator and the federal courts.

The NPDES permits contain three major parts: effluent limitations,
compliance schedules, and monitoring and reporting requirements. Generally, each pollutant that is regulated in a permit has effluent limitations
expressed in terms of mass or load (a measurement of quantity) and concentration (a measurement of quality) and its mass and concentration limitations each have a maximum daily discharge limitation and an average
monthly discharge limitation. For example, the permittee's discharge of total
suspended solids might have an average monthly mass discharge limitation of 27 kilograms per day (kgd), a maximum daily discharge limitation
of 185 kg, an average monthly concentration discharge limitation of 30 milligrams per liter (mgL) and a maximum daily discharge limitation of 100 mgL.
These numbers are based on categorical technology-based effluent limitations, more stringent water quality-based limitations, or best professional
judgment where technology-based limitations have not yet been promulgated or the discharger has received a variance.

This method of expressing effluent limitations in permits is referred
to as the "chemical-specific" approach to regulation. However, where mixed
waste streams are concerned, effluent limitations expressed in terms of
bioassays are becoming increasingly common for both industrial and
municipal dischargers.

> In the most widely used procedure, permit writers calculate the amount of
> flow in a receiving water at low flow conditions (such as the lowest one week
> average flow that a statistical analysis of stream flow data indicates will occur
> once every 10 years). Permitting authorities then use a simple formula to de-
> termine allowable effluent toxicity: the instream concentration of the effluent
> in the receiving water during the receiving water's low flow periods must not
> be greater than the concentration at which the effluent causes unacceptable
> toxic effects on the most sensitive test species. This simple "wasteload alloca-
> tion" is used as the target for effluent treatment and a permit limit is calculat-
> ed which ensures that the target wasteload allocation for effluent toxicity is
> not exceeded instream.[19]

Once again, it must be emphasized that bioassay-based limitations may
be insufficient where the effluent causes long-term ecosystem effects or contains substances that are human carcinogens or bioaccumulative in fish. The

EPA is cautioning that waterbodies used for potable water and sportfishing be protected by single chemical effluent limits.

There is one situation where "best management practices" (BMPs) for a nonpoint source can be included in a discharge permit. Where a category of industrial point sources discharges a toxic pollutant, the EPA is authorized to include within its categorical effluent limitations BMPs "to control plant site runoff, spillage or leaks, sludge or waste disposal, and drainage from raw material storage which . . . are associated with or ancillary to the industrial manufacturing or treatment process . . . and may contribute significant amounts of such pollutants to navigable waters."

Compliance schedules set interim requirements to assure that the compliance date is met. Interim requirements may include retaining a consultant, contracting for installation of equipment, installing equipment on schedule, and attaining interim effluent limitations. Each interim requirement is enforceable independently. Dischargers of toxics are frequently required to comply with toxicity control plans containing implementation schedules.

Before and after compliance, a permittee must carry out extensive self-monitoring and reporting. A permit specifies how, when, and where monitoring must be conducted. As of July 1986, 38% of major industrial permits and 10% of major municipal permits contained bioassay monitoring requirements.[19] A permit violation is referred to as an "excursion." Serious excursions must be reported within 24 hours, while less serious excursions must be reported on a monthly "Discharge Monitoring Report" (DMR).

Certain excursions—called "upsets" and "bypasses"—are defenses to some enforcement actions. An "upset" is an exceptional incident in which there is an unintentional and temporary noncompliance with technology-based effluent limitations because of factors beyond the reasonable control of the permittee. It cannot be caused by operational error, improperly designed treatment facilities, lack of preventive maintenance, or improper operation. An upset is not a defense to a violation of water quality-based effluent limitations. Neither is a "bypass," which is defined as an "intentional diversion of waste streams from any portion of the treatment facility." Bypasses that do not cause technology-based effluent limitations to be exceeded are permissible in order to perform "essential maintenance to assure efficient operation." Bypasses that exceed these limitations are allowable if they are "unavoidable to prevent loss of life, personal injury, or severe property damage."

Discharge permits may be granted for a maximum of five years. Congress has consistently refused EPA requests to authorize permits of longer duration. When a permit expires, the permittee must apply for a renewal. Assuming timely renewal application, the old permit remains in effect until the renewal is granted. Once a new permit is issued, a discharger is bound by it even if he is contesting it in the permitting agency or in court. Permits may be modified or revoked and reissued during the permit period. Common grounds for modification or revocation and reissuance are alterations to permitted facilities, new information, supervening federal regulations, unforeseen events, and procurement of a variance. Permits may also be terminated before the scheduled expiration date for nondisclosure, noncompliance, or alleviation of an emergency situation.

The 1987 CWA amendments place limitations on the issuance of new permits that are less stringent than their predecessors ("anti-backsliding"). Permits based on Best Professional Judgment (BPJ) or water quality may not be renewed, reissued, or modified to contain less stringent effluent limitations except where: (a) material and substantial alterations or additions to the permitted facility occurred after permit issuance, justifying a less stringent limit; (b) the EPA determines that technical mistakes or mistaken interpretations of law were made in issuing the permit. (This ground applies only to BPJ effluent limitations, not to water quality-based effluent limitations.); (c) a less stringent effluent limitation is necessary because of events over which the permittee has no control and for which there is no reasonably available remedy; (d) the permittee has received another permit variance or modification under the CWA; or (e) the permittee has installed and properly operated and maintained the treatment equipment necessary to meet the permit limits, but has been unable to meet the limits.

However, in no event can the revised effluent limitation be less stringent than an applicable technology-based effluent limitation or lead to a violation of water quality standards.

Another subsection of the "anti-backsliding" amendment could, nevertheless, encourage backsliding. Wasteload allocations and TMDLs may be revised to be less stringent as long as water quality standards are not violated. Thus, where a waterbody becomes appreciably cleaner because of the control of nonpoint sources of pollution, point sources that had installed more sophisticated technology in order to meet their water quality-based limitations could obtain a reallocation and abandon the new waste treatment methods.

Thirty-seven states are currently administering NPDES permit programs under agreements with the EPA. This number may well increase because, under the 1987 amendments, the EPA may, for the first time, make partial delegations to a state. The EPA may delegate to a state part of the NPDES program where the state partial permit program covers, at a minimum, either administration of a major category of discharges or a major component of the state's permit program. A state may return, or the EPA may withdraw, such partial delegation. However, this does not authorize a state to return parts of a program approved prior to enactment of the amendments.

Basically, in order to obtain delegation of the NPDES program from the EPA, a state must convince the EPA that its program is consistent with the CWA and EPA regulations, and that the state possesses the resources and statutory authority to implement it. The state issues future discharge permits once having obtained EPA program approval. State discharge permits are final, subject only to EPA veto of a permit as being "outside the guidelines and requirements" of the CWA. Conditions in a state-issued permit may not be less stringent than EPA regulations (for example, the EPA's categorical effluent limitations) but may be more stringent. If the state does not adequately administer the permit program, the EPA may rescind program approval and resume primary enforcement responsibility. A state can also lose its permit program for failure to develop an acceptable pretreatment program.

The danger of this "cooperative federalism" is that program goals will be compromised by states, which either lack the resources to administer the program effectively or remain more vulnerable to political pressure by dischargers than the EPA might be. This danger is heightened by inadequate funding for state program grants and EPA supervision. Without adequate resources, the EPA cannot accurately gauge the strength of a state program or credibly threaten to rescind it. In fact, states have gained leverage against the EPA by threatening to "return" their programs. Serious doubts have been raised about the quality of many state programs. For example, environmental groups have sued the EPA to rescind programs in Michigan and Wisconsin, and have petitioned the EPA to review programs in Kentucky and Virginia. The EPA itself has warned nine states about permit backlogs. Colorado and the EPA argued for years about whether the board that issued permits in Colorado might include industry representatives.

Where the EPA is still the permitting authority, a state must be afforded an opportunity to "certify" to more stringent permit conditions. If the state denies certification, the permit may not be issued, and the more stringent conditions of a state certification are binding on the EPA. If the state waives certification, or fails to certify within a reasonable time, the EPA is free to issue the permit as drafted. State certification decisions, as well as all state permit decisions where the state is the permitting authority, must be contested in state administrative tribunals and state courts.

Section 313 of the CWA waives federal sovereign immunity in the water pollution control area. Federal facilities and the federal personnel operating them must comply with all federal, state, interstate, and local water pollution control requirements "in the same manner, and to the same extent as any nongovernmental entity including the payment of reasonable service charges." This means that federal facilities must obtain discharge permits, comply with effluent limitations and reporting requirements, and respond to enforcement just as any other discharger. Innovative or alternative wastewater treatment must be used in new construction at federal facilities unless the cost of innovative treatment exceeds the life cycle cost of the most cost-effective alternative by more than 15%.

There are, however, some potential variances for federal facilities and their operators. The president may grant a variance from CWA requirements to any federal facility if he determines it to be in the "paramount interest of the United States" to do so. No variance can be granted to a new source, an indirect discharger, or a discharger of toxic pollutants. Variances run for one year, subject to an unspecified number of one-year extensions. In addition to variances for particular dischargers, the president may also issue variances to "any weaponry, equipment, aircraft, vessels, vehicles, or other classes or categories of property . . . which are uniquely military in nature." Exemptions for classes of military property may run for three years, subject to extension. Finally, the EPA may waive the innovative and alternative treatment requirement "in the public interest."

Chapter 32
THE CLEAN WATER ACT: ENFORCEMENT

Enforcement credibility is critical to the success of any regulatory strategy for controlling water pollution. Prior to 1972, federal water quality statutes were inherently unenforceable, and little progress was made in cleaning up America's waterbodies.[4] The CWA of 1972 provided enforceable duties and a new arsenal of enforcement sanctions to deter potential violators and punish illegal conduct. Any remaining deficiencies in the EPA's enforcement powers were remedied by the 1987 amendments. The question is no longer whether the CWA is enforceable, but to what extent its powerful enforcement weapons are being used.

The Environmental Protection Agency has been given broad powers of inspection, monitoring, and entry of discharger premises in implementing the CWA. A discharger can be required to keep records, make reports, install and maintain monitoring equipment, sample in accordance with EPA instructions, and provide information on request. Moreover, the EPA has the right to enter discharger premises and, at reasonable times, inspect and copy records, inspect monitoring equipment, and sample effluent. Because of recent federal court decisions qualifying administrative rights of entry under federal regulatory statutes, the EPA routinely obtains a search warrant whenever the discharger refuses entry. All EPA records, including discharge monitoring reports, are available to the public unless the discharger demonstrates that the information should be treated as a confidential trade secret; but effluent data cannot be kept confidential.

The most important thing to remember about enforcement is that it is discretionary. Whether, when, and how to enforce are decisions to be made by the enforcement authority. These decisions are further complicated where, as in water pollution control law, a state and the federal government have concurrent enforcement power. "Who will enforce?" becomes the first order of business.

Where a state is administering an approved permit program, it is the primary enforcement authority. Its enforcement powers arise from state law. The EPA has the authority to immediately bring enforcement action in an administering state, but it generally does not exercise it. Instead, where the EPA finds a violation in an administering state, it either contacts the state informally or issues a 30-day notice of violation to the discharger and the state. If the state does not take "appropriate enforcement action" within the 30-day period, the EPA can pursue its own enforcement. When a state is apparently neglecting its entire enforcement program, the EPA can give notice to the state and assume enforcement of state-issued permits until the state rectifies the situation.

GOVERNMENTAL ENFORCEMENT REMEDIES

The EPA's enforcement remedies are administrative compliance orders, administrative fines, civil suits, and criminal actions. These remedies need not be sought in any particular order, but in practice criminal sanctions are a last resort.

Until the 1987 amendments, the EPA lacked authority to levy administrative fines. This deficiency was a major obstacle to enforcement, because the EPA was forced to bring a time-consuming civil action whenever a compliance order was violated. Now there is a two-tiered administrative penalty system. Class I penalties may not exceed $10,000 per violation, with a maximum penalty of $25,000. Informal hearings are available in Class I penalty situations. Class II penalties may not exceed $10,000 per day of violation, with a maximum penalty of $125,000. There is an opportunity for a formal, trial-type hearing in Class II penalty situations. The EPA can determine in a specific case which class of penalties to apply. Public notice and an opportunity to comment must be provided before an administrative penalty is assessed.

Where the violation is more serious, the EPA may bring a civil action for an injunction and a civil fine of up to $25,000 per day of violation. In determining the amount of a civil penalty, the court must consider the seriousness of the violation, any economic benefit to the violator, past violations, good faith efforts to comply, the economic impact of the penalty on the violator, and other relevant factors. The U.S. Supreme Court has held that an injunction need not be granted automatically when the CWA is violated. A court may "balance the equities" and determine whether an injunction would be in the public interest.[20]

Any person, including a "responsible corporate officer," who "willfully or negligently" violates the CWA is subject to criminal prosecution. An unauthorized discharge of a hazardous substance into a sewer system, which foreseeably could cause personal injury or property damage, is now a criminal offense. Negligent violations are punishable by fines of between $2,500 and $25,000 per day of violation, or by one year in jail, or both. These maximum sanctions are doubled for second and subsequent offenders. Knowing violations are punishable by a fine of between $5,000 and $50,000 per day of violation, or by up to three years in prison, or both. Once again, potential penalties are doubled for repeat offenders. A person who is found guilty of "knowing endangerment" (i.e., knowingly placing another person in imminent danger of death or serious bodily injury) is subject to fines of up to $250,000 and imprisonment for up to 15 years, with a possible doubling for repeat offenders. Persons making false statements in applications or reports, or tampering with monitoring devices, can be fined up to $10,000 and imprisoned for up to six months. Because of the difficulty of obtaining criminal convictions, the criminal sanction is reserved for repeated and heinous cases of water pollution. Although very much the exception, jail sentences for water pollution are being imposed with increasing frequency.

Thus far, enforcement options under section 309 of the CWA have been discussed, but other parts of the act also have a bearing on enforcement. Section 504 authorizes the EPA to immediately sue for an injunction "or to take such other action as may be necessary" where a discharge "is presenting an imminent and substantial endangerment to the health . . . or to the welfare of persons." Section 508 prohibits federal agencies from contracting with, or making loans or grants to, listed violators of the CWA. This potentially powerful sanction, however, has only been used in a few cases. Section 401(h) authorizes the EPA or a state with an approved program to seek a court order barring new indirect discharges where a POTW is overloaded and violating its permit. These orders, called "sewer bans," are prominent features of many state enforcement efforts, but the EPA has generally avoided them.

CITIZEN SUITS

Citizen suits are vital adjuncts to governmental enforcement of the CWA. Section 505 of the CWA gives any citizen standing to bring a civil action in federal district court against a discharger currently violating an effluent limitation or compliance order, or against the EPA for failure to perform a nondiscretionary duty. The EPA cannot be sued to compel enforcement because enforcement is discretionary. A court can, if it is in the public interest, grant an injunction in a citizen suit. It can also impose a civil fine on a violating permittee, with the fine being paid into the federal treasury. Citizen suits are frequently settled by court-approved consent decree, with the defendant making payments to environmentally beneficial projects (e.g., university research and public education). Damages cannot be recovered in a citizen suit, but an individual claim for damages may be joined in one lawsuit with a citizen suit.

No citizen suit can be brought unless a 60-day notice of intent has been given to the EPA, the relevant state, and the alleged violator, or to the EPA alone if it is the defendant. The 60-day waiting period is waived in suits against new sources and dischargers of toxics. A citizen suit is barred if the EPA or the state is "diligently prosecuting" a civil or criminal action to enforce the standard or order, or if an administrative penalty has been recovered. But if a citizen suit is barred on this ground, the citizen has a right to "intervene" (be included as a plaintiff) in the governmental lawsuit. Thus, the 60-day notice and citizen suit can be used to stimulate governmental agencies to bring new enforcement actions and proceed seriously with existing actions.

One incentive for meritorious citizen suits is that the court may award costs of litigation, including reasonable attorney and expert witness fees, to any party (plaintiff or defendant) who substantially prevails in the suit.

Citizen plaintiffs have won or settled dozens of CWA citizen suits, with single fines and settlements reaching into the millions of dollars. In fact, citizen suits against current permit violators are comparatively easy to win. A discharger's own discharge monitoring reports are admissible as evidence because they are formal records required to be kept in the ordinary course of business. In most instances, the defendant's DMR is conclusive proof of a violation, and plaintiff is entitled to "summary judgment," without going to trial, on the issue of liability. Witnesses are only called to determine the amount of the fine.

EPA ENFORCEMENT

Between 1972 and 1985, the EPA's enforcement record was poor, especially with regard to POTWs. The GAO reports in 1978 and 1983[21],[22] indicated that permittees were reporting a high degree of noncompliance with permit conditions in their discharge monitoring reports to the EPA. Some instances of noncompliance involved significant violations of permit limitations for toxic substances. When the EPA responded at all to these violations reported by the permittees themselves, its response was neither timely nor strong in most cases. The General Accounting Office recommended a thorough reevaluation of the EPA's enforcement system. Apparently, the EPA performed this reevaluation, because its enforcement program has been substantially improved in recent years. Most important, the EPA has finally begun enforcement proceedings against POTWs violating their permits. Fines have been levied against such major cities as Boston, the District of Columbia, Philadelphia, and Los Angeles. These and other POTWs have been placed on compliance schedules that will bring them into compliance by 1988 or as soon thereafter as possible. Moreover, the EPA has enforced, in some cases criminally, against indirect dischargers who discharge incompatible toxic pollutants into POTWs. Now that the EPA has the authority to levy administrative fines, its enforcement system should be even more effective.

However, there is no evidence that the EPA has responded to the GAO's criticism that state-managed permit programs are not being adequately supervised. With 37 states administering the NPDES permit system, this may be a major problem.

Chapter 33

THE CLEAN WATER ACT: INTERSTATE WATER POLLUTION

Some of our most heavily polluted waterbodies lie within the jurisdictions of two or more states. With the Clean Water Act administered primarily at the state level, difficult questions arise about whether a state can impose permit conditions on, or enforce against, dischargers in another state. As with water allocation, interstate organizations have been formed to deal with the interstate aspects of the system.

INTERSTATE WATER POLLUTION CONTROL COMPACTS

Interstate water pollution control compacts (IWPCCs) are beset by the same political viability problems as are other regional institutions (see Chapter 12). For this reason, Congress has entrusted implementation of the CWA to the EPA in cooperation with the states and Indian tribes.

However, IWPCCs do play a role in America's water pollution control effort. There are about a dozen interstate compacts that are involved with water pollution control. Some of these are: Bi-State Metropolitan Development District Compact (Missouri and Illinois); Delaware River Basin Compact (Delaware, New Jersey, New York, Pennsylvania, and the United States); Great Lakes Basin Compact (Illinois, Indiana, Michigan, Minnesota, New York, Ohio, Pennsylvania, and Wisconsin); Klamath River Basin Compact (California and Oregon); New England Interstate Water Pollution Control Compact (Connecticut, Maine, Massachusetts, New Hampshire, Rhode Island, and Vermont); New York Harbor (Tri-State) Interstate Sanitation Compact (New York, New Jersey, and Connecticut); Ohio River Valley Water Sanitation Compact (Illinois, Indiana, Kentucky, New York, Ohio, Pennsylvania, Tennessee, and West Virginia); Potomac Valley Conservancy District Compact (District of Columbia, Maryland, Pennsylvania, Virginia, and West Virginia); Susquehanna River Basin Compact (New York, Maryland, Pennsylvania, and the United States); and the Tennessee River Basin Water Pollution Control Compact (Alabama, Kentucky, Mississippi, and Tennessee). Each compact established an interstate water pollution control compact commission made up of representatives of the signatories. On the international level, the Great Lakes Water Quality Agreement (1972, 1978, 1983) between the United States and Canada authorizes the International Joint Commission to perform pollution control functions in addition to its water diversion functions mentioned in Part I.

All of these compacts were formed before 1972, and the subsequent CWA placed them in a subordinate position. Section 101 of the act declares the policy of Congress "to recognize, preserve, and protect the primary rights of States to prevent, reduce, and eliminate pollution." Congress avoided creating an additional bureaucratic layer in water pollution control, perhaps because compacts cover a relatively small portion of the United States. Commissions do not administer construction grants programs or issue discharge permits, although they provide input to the EPA and states that administer these programs.[23] Where commissions possess enforcement powers they exercise them sparingly, because they are cautious about intruding on state prerogatives. Similarly, under the Great Lakes Water Quality Agreement, the International Joint Commission can set standards and perform monitoring, but enforcement is left to the two nations themselves. Although some commissions have authority to regulate land use in their basins, they concentrate on large, individual projects and do not conduct comprehensive programs of nonpoint source pollution control. Finally, the basinwide planning carried out by commissions is on a larger scale than the CWA's water quality management planning process.

Either exclusively or jointly with basin states, most commissions set water quality standards, perform water quality monitoring, undertake research projects, and especially do modeling and wasteload allocations on interstate waters.[23] Perhaps most importantly, they provide a coordinating mechanism and a forum for the exchange of information among basin states.

Existing compacts will probably survive despite their minor, somewhat duplicative role under the CWA. In the first place, a number of them (for example, the Delaware and Susquehanna commissions) perform other functions such as flood control and water supply management. Second, commissions are so helpful to basin states as negotiation and communication facilitators that the states continue to fund them with state and passthrough federal monies. In short, existing compacts are functional and established. However, given the CWA's structure, it is improbable that new ones will be formed. One possibility, nevertheless, is that interstate water supply commissions in the West will be given water pollution control functions by their constituent states.

INTERSTATE PERMITTING AND ENFORCEMENT

The legal principles applicable to this area can best be illustrated by reference to a recent United States Supreme Court decision.

In *International Paper Company* v. *Oulette*,[24] plaintiffs were property owners on the Vermont shore of Lake Champlain, which forms part of the border between Vermont and New York. Defendant International Paper Company (IPC) operates a paper mill on the New York side of the lake. Plaintiffs sued in federal court under Vermont common law, demanding an injunction and damages on the ground that defendant's discharges in New York constituted a nuisance in Vermont. Defendant was operating under a discharge permit issued by a New York agency.

Until 1981, an interstate pollution control action could have been predicated on the so-called "federal common law of nuisance," at least where there was a governmental plaintiff. This doctrine gave federal courts the authority to impose more stringent standards, under the vague criteria of federal common law, than those imposed by the Clean Water Act. But in 1981 the United States Supreme Court abolished the federal common law of nuisance doctrine in water pollution cases.

The *Oulette* case involved the question of whether an affected state may impose its own law on dischargers located in another state. The Supreme Court answered this question in the negative:

> While source States have a strong voice in regulating their own pollution, the CWA contemplates a much lesser role for States that share an interstate waterway with the source (the affected States). Even though it may be harmed by the discharges, an affected State only has an advisory role in regulating pollution that originates beyond its borders. Before a federal permit may be issued, each affected State is given notice and the opportunity to object to the proposed standards at a public hearing. An affected State has similar rights to be consulted before the source State issues its own permit; the source State must send notification, and must consider the objections and recommendations submitted by other States before taking action. Significantly, however, an affected State does not have the authority to block the issuance of the permit if it is dissatisfied with the proposed standards. An affected State's only resource is to apply to the EPA Administrator, who then has the discretion to disapprove the permit if he concludes that the discharges will have an undue

impact on interstate waters. Also, an affected State may not establish a separate permit system to regulate an out-of-state source. . . . Thus, the Act makes it clear that the affected states occupy a subordinate position to source States in the federal regulatory program.

By extrapolation, common law actions under the common law of the affected state against dischargers in the source state are also preempted by the Clean Water Act.

The Vermont property owners have two legal remedies. If IPC is in violation of its New York discharge permit, they could bring a CWA citizen suit against it. Whether or not there is a permit violation, they could sue under New York common law because, under the CWA, a source state may set as stringent discharge conditions as it sees fit, and this power extends to court-imposed conditions.

Chapter 34
THE CLEAN WATER ACT: NONPOINT SOURCE CONTROL

The drafters of the CWA considered point sources to be not only the primary causes of water pollution but also the most easily regulated. Therefore, they emphasized the construction and regulation of publicly owned treatment works and the issuance to them and to industrial dischargers of permits embodying technology-based effluent limitations—two programs specifically aimed at point source control. Early implementation activities reflected this emphasis.[25] Since 1972, however, nonpoint sources have come to be recognized as the major contributors to many waterbodies of such pollutants as biochemical oxygen demand, suspended solids, and toxics such as pesticides from agricultural runoff and lead from urban stormwater runoff. A survey conducted by the Association of State and Interstate Water Pollution Control Administrators in 1984 indicated that 78% of the states saw their nonpoint source problems as greater than or equal to those caused by point sources.

Nonpoint sources are diffuse sources of water pollution that do not discharge at single locations (e.g., pipes or ditches) but whose pollutants are generally carried over or through the soil and ground cover by way of stormflow processes. Major sources of nonpoint pollution are agriculture and silviculture, urban stormwater, mineral extraction, residential construction, disposal of residual wastes, and storage of industrial materials.

For the most part, technology-based effluent limitations are less effective in controlling nonpoint sources than the nontechnological (or "low-tech") approaches of land use controls and land management practices. These approaches—referred to as "Best Management Practices" (BMPs)—include methods, measures, or practices to prevent or reduce water pollution, consisting of structural and nonstructural controls and operation and maintenance procedures. Usually, BMPs are applied as a system of practices rather than a single practice. BMPs are selected on the basis of site-specific conditions that reflect natural background and political, social, economic, and technical feasibility. For example, a set of BMPs to reduce runoff of herbicides and pesticides from a particular farm might include reduced applications of chemicals, contour plowing, and vegetated stream buffers.

Because nonpoint source control is inherently site-specific and land-use oriented, Congress, in 1972, reserved it to state and local governments. It was felt that BMPs should be developed and implemented by the level of government closest to the source of the problem, the level of government authorized to make land-use decisions. Congress saw no need for a national regulatory program as with point sources.

Thus, the Clean Water Act of 1972 dealt with nonpoint source control in sections devoted primarily to planning. This has led to an enforceability problem for nonpoint source control programs.

SECTION 208

The term "water quality management planning" refers specifically to sections 208, 303(e), and 106(c) of the CWA, as interpreted by federal planning regulations. However, where nonpoint sources are concerned, section 208 is the focal point.

Entitled "Areawide Waste Treatment Management," section 208 established a procedure whereby states or designated regional agencies were required to formulate and implement strategies to control both point and nonpoint source pollution. The relationship of section 208 to point source control is discussed in Chapter 35. The mandate for this type of planning can be found in subsection 101(a)(5) of the act, which sets forth the "national policy that areawide waste treatment management planning processes be developed and implemented to assure adequate control of sources of pollutants in each State." Consistent with this policy, section 201 of the CWA further directs: "To the extent practicable, waste treatment management shall be on an areawide basis and provide control or treatment of all point and nonpoint sources of pollution, including in place or accumulated pollution sources."

Under section 208, the governor of each state was required to designate areas with "substantial water quality control problems" and to name

for each area "a single representative organization, including elected officials from local government or their designees." By and large, 208 planning
agencies have been councils of government or county governmental units.
Each of these organizations or its planning agency was to then formulate
and institute "a continuing areawide waste treatment management planning process" which "shall contain alternatives" for waste treatment
management applicable to "all wastes generated within the area involved."
Federal planning grants provided funds for planning studies and the planning process. Plans prepared under this process were to be submitted for
certification by the governor and approval by the EPA. Section 208 provided for interstate cooperation where the relevant area lay in more than one
state, and authorized local officials to form their own planning agency if
the governor failed to act. Finally, the state itself was the planning agency
for all portions of the state not designated by the governor or local authorities as areawide waste treatment management areas.

In its water quality management plan, each planning agency had to
identify and recommend for designation by the governor "one or more waste
treatment management agencies (which could be an existing or newly created
local, regional, or State agency or political subdivision)" to implement the
plan. The EPA could reject the governor's designation of a management
agency if the designated agency did not have authority to carry out the plan.

Water quality management plans had to be submitted to the EPA within
three years after receipt of the EPA's planning grant. Over 200 state and
areawide plans have been completed. Water quality management planning
for nonpoint sources required the identification of nonpoint sources of pollution and the development of "procedures and measures (including land
use requirements)" for controlling them "to the extent feasible." Each plan
had to address itself to the following nonpoint sources where they were relevant: agricultural and silvicultural nonpoint sources; mine-related sources;
construction-related sources; saltwater intrusion resulting from reduction
of freshwater flow; disposal of residual wastes; disposal of wastes on land
or in subsurface excavations; hydrologic modifications due to construction
of dams and other flow diversion mechanisms; and urban stormwater runoff.

Two aspects of this list deserve attention. First, the protection of
groundwater quality is mentioned with regard to several of these sources.
Section 208 is the only major part of the CWA that specifically relates to
groundwater. Thus, it has served as a useful instrument for conjunctive
management of groundwater and surface water resources. Second, section
208 recognizes relationships among water diversions, water resources
development, and water quality. In this sense it has enhanced comprehensive water resources management.

Nonpoint source control under section 208 featured BMPs established
by planning agencies and implemented by management agencies designated
in the plan. Water quality management plans contain ranges of BMPs for

varieties of nonpoint sources. Management agencies then choose the mix of BMPs applicable to specific sites and implement site-specific strategies. Thus, the nonpoint source control element of a plan was, by its very nature, an abstract study rather than a set of enforceable requirements.

Furthermore, planning agencies were not required to develop regulatory programs for implementing BMPs. Subsidies, public education, institutional coordination, technical assistance, and voluntary compliance programs were also acceptable to the EPA. Many planning agencies recognized early in the process that new land-use controls would be controversial. Thus, one encounters few recommendations for new regulatory programs in the plans.

The unenforceability of the water quality management plans themselves, and their failure to produce recommendations for enforceable regulatory programs, might lead some to believe that the nonpoint source component of section 208 has been a complete failure. There has been no comprehensive evaluation of section 208. But this author concludes that water quality management planning has generated significant, although sporadic, results.

In New Jersey, the money invested in water quality management planning has paid dividends at the state level and in a number of counties. Statewide, regulatory programs for control of soil erosion and sedimentation from construction projects, stormwater runoff, groundwater contamination, hazardous waste site leachate, and spills have been inaugurated or substantially strengthened because of information developed during the 208 process. This information has also led to state-financed studies of watershed management, groundwater contamination, stream corridor protection, and stormwater runoff. Strong county programs, based on the 208 plans, have been initiated in stream monitoring, stormwater control, groundwater protection, septic system control, and public education. On the negative side—and this appears to be the case nationally—virtually nothing has been accomplished in reducing urban stormwater or agricultural nonpoint source pollution. Moreover, other counties have been unreceptive to implementing nonpoint source controls.

In North Carolina, 208 has been the catalyst for protection of two reservoirs in a rapidly developing area.

> The Triangle J Council of Governments, the regional planning agency for the Research Triangle Area, has developed technical expertise in water quality management through its experience as a former 208 agency, and has assumed a regional leadership role in water resource issues. However, the COG lacked the regulatory authority to implement a watershed program. The Department of National Resources recognized the importance of water quality to the continued economic success of the Research Triangle area and joined forces with Triangle J to develop a comprehensive watershed protection plan. [They] formed a Steering Committee composed of the chief elected officials from each local government in the two watersheds. The state used as leverage its permit authority over wastewater treatment plants to gain the cooperation of all local

governments in the two watersheds. Local leaders emphasized the need to "share the burden equally" through the use of uniform, but flexible, standards. They also demanded that the recommendations be technically based. The result of an intense effort was the "State-Local Action Plan."[26]

In the plan, the state committed itself to expand monitoring for toxic chemicals, require all new, and some existing, POTWs to remove phosphorus, and establish a $2 million cost-share program for use of agricultural BMPs. Local governments undertook to implement watershed zoning, stream buffers, stormwater management, sedimentation and erosion control, and restrictions on impervious cover. "These recommendations have been incorporated into local zoning and subdivision regulations by many local governments in the two watersheds."[26]

Water quality management planning was the incubator for an agreement, signed in 1983, by the Governors of Virginia, Maryland, and Pennsylvania, the Mayor of Washington, D.C., and the EPA Administrator to control the nonpoint pollution that was destroying Chesapeake Bay. In the Chesapeake Bay Agreement, the signatories agreed to appropriate funds and enact legislation to control nonpoint sources within their jurisdictions, and in the 1987 CWA amendments $10 million per year was authorized for matching grants to states to implement the Agreement. Maryland has enacted a Chesapeake Bay Critical Area Law, establishing a commission to develop regulations for a 1000-foot buffer strip around the Bay's shoreline and along its tidal tributaries. Pennsylvania has established a cost-share program with farmers along the Susquehanna River to reduce nitrogen loads from agriculture.

The Chesapeake Bay model has developed into a national program—the National Estuary Program. In the 1987 amendments, $12 million per year through 1991 has been authorized for matching grants to fund management conferences to prepare management plans for estuaries and then to fund states to implement the plans. These funds can be augmented by state contributions from their state revolving funds and EPA payments from reserved funds. Management conferences will also review federal financial assistance for consistency with the plans.

Another federal program directed toward controlling nonpoint source pollution is the Clean Lakes Program: $30 million per year through 1990 has been authorized for matching grants to states, and a $40 million Clean Lakes Demonstration Program has been authorized.

The New Jersey, North Carolina, and Chesapeake Bay examples illustrate section 208 at its best. In these instances, the planning process identified problems, raised the public's consciousness of nonpoint source problems, and brought people together to develop mutually acceptable solutions. Where 208 has worked, it has served as an information-gathering, public education, facilitation, and consensus-building mechanism.

Nevertheless, 208 has failed more often than it has succeeded. This overall failure was caused primarily by 208's naive regionalist philosophy. In reality, local governments were reluctant to relinquish planning and land management authority to areawide agencies whose goal of improved water quality might clash with their goal of development and growth. For the areawide planning agencies, lacking enforcement powers and a mandate to produce an enforceable plan, the price of gaining local acceptance was too often a vacuous plan. Moreover, many state officials perceived areawide planning and management agencies as potential rivals. Thus, the fledgling areawide agencies had only a tentative ally in the EPA (an agency that has always avoided being perceived as a land-use regulatory agency), and few allies at the state and local levels. Fundamentally, 208 planning was a massive public education program. And like all public education efforts, it achieved reform only where the public could be taught to desire the change.

THE 1987 AMENDMENTS

The 1987 amendments add a new section 319—entitled "Nonpoint Source Management Programs"—to the Clean Water Act. As with section 208 of the 1972 act, Congress has relied on the grants approach, rather than a federal regulatory approach, to control nonpoint sources. The regional approach has been abandoned. States have replaced areawide agencies as the nerve centers of nonpoint source control. States are now only required to develop programs "in cooperation with" areawide agencies, and regional bodies can only receive federal funding if a state does not submit an adequate management program to the EPA.

Within 18 months of enactment, each state must submit to the EPA Administrator a two-part document containing (1) a State Assessment Report, and (2) a State Management Program. The Report must:

(a) identify those surface waters within the state, which, without additional action to control nonpoint sources of pollution, cannot reasonably be expected to attain or maintain applicable water quality standards or other Clean Water Act goals or requirements;

(b) identify those categories and subcategories of nonpoint sources or, where appropriate, particular nonpoint sources that add significant pollution to the waters identified in (a);

(c) describe the process, including intergovernmental coordination and public participation, for identifying Best Management Practices (BMPs) and measures to control each category and subcategory of nonpoint sources, and, where appropriate, particular nonpoint sources and to reduce, to the maximum extent practicable, resulting nonpoint source pollution; and

(d) identify and describe state and local programs to control nonpoint source pollution in problem areas.

The Management Program, "which such state proposes to implement in the first four fiscal years beginning after the date of submission," must

(a) identify BMPs and measures that will be undertaken to reduce nonpoint source loadings in problem areas "taking into account the impact of the practices on ground water quality,"

(b) identify programs "(including, as appropriate, nonregulatory or regulatory programs for enforcement, technical assistance, financial assistance, education, training, technology transfer, and demonstration projects)" to achieve implementation of BMPs,

(c) include a schedule containing annual milestones for utilizing program implementation methods identified in (b) and implementing BMPs; the schedule must "provide for utilization of [BMPs] at the earliest practicable date,"

(d) include a certification by an appropriate state legal official that the state possesses, or will seek as expeditiously as practicable, adequate authority to implement the management program,

(e) identify sources of funding, other than section 319 grants, for management program implementation, and

(f) identify a process for reviewing federal assistance or development projects in terms of their "consistency" with the state nonpoint source management program.

In developing and implementing a management program, a state shall, to the maximum extent practicable, use existing public and private organizations that have expertise in control of nonpoint sources of pollution. Moreover, management programs must, to the maximum extent practicable, be developed and implemented on a watershed-by-watershed basis.

The Administrator must approve or disapprove the state submissions within 180 days or they are deemed approved. The Administrator may disapprove a program or portion of it upon determination, among other considerations, that it is not likely to satisfy the goals and requirements of the CWA, or that the practices and measures proposed in the plan are not adequate to reduce nonpoint source pollution and to improve water quality. The state shall have three months to revise its plan and the Administrator shall approve or disapprove the revised program within three months. If a state fails to submit the report, or if it is not approved, a local public agency or organization with expertise in and authority to control nonpoint sources may, with the approval of the state, develop and implement a program for its area.

Where waters in a state with an approved program are not meeting applicable water quality standards or the goals and requirements of the CWA because of upstream pollution, the state may petition the EPA to convene, or the EPA may initiate, an interstate management conference to develop an agreement. No agreement can supersede or abrogate water rights established by interstate water compacts, Supreme Court decrees, or state water laws. Nor shall any agreement apply to pollution subject to the Colorado River Basin Salinity Control Act. If states reach agreement through the conference, their management programs will be revised to incorporate the agreement.

Once a state's report and management program have been approved, it becomes eligible for a federal matching grant of 60% at maximum. Federal grant funds must be used for implementation of the management program, and $400 million has been authorized over four years for these grants. Grants will be made on an annual basis, and no subsequent grant will be made to a state unless the state has "made satisfactory progress in the preceding fiscal year in meeting the schedule specified by such State. . . ." Grant funds may be applied to cost-sharing with individuals only for demonstration projects. No state may receive a grant unless it promises to maintain its other nonpoint source program funding at or above its average level for the two years preceding enactment. In making grants, the EPA may give priority to effective mechanisms that will (1) control particularly difficult nonpoint source problems, (2) implement innovative methods or practices, (3) control interstate nonpoint source pollution, or (4) carry out groundwater quality protection activities related to nonpoint source pollution. A separate $7.5 million is authorized for groundwater quality protection grants.

Section 319 solves one of section 208's problems—the "areawide agency problem"—but the "enforceability problem" remains. Will the threat of losing federal grant funds persuade states to impose the politically unpalatable land use controls that will be necessary to significantly reduce nonpoint source loadings? Or is it time for a federal regulatory program?

Chapter 35

THE CLEAN WATER ACT: 208 POINT SOURCE PLANNING

Section 208 plans retain continuing validity mainly as devices for planning POTWs and assuring that they do not become overloaded.

INDUSTRIAL POINT SOURCE PLANNING

Originally, Congress attempted to integrate industrial discharge permit issuance with water quality management planning by prohibiting the issuance of a discharge permit which was "in conflict" with an approved plan. Theoretically, water quality management plans were to provide for the maintenance of effluent-limited stretches and set wasteload allocations and water quality-based effluent limitations on water quality-limited stretches. However, this aspect of 208 planning became a dead letter because areawide planning agencies did not possess the technical resources to cope with prescribing industrial control technology or performing water quality modeling and wasteload allocations. One of 208's major deficiencies was the unrealistically short three-year period allowed for plan preparation. In order to meet this deadline, the EPA required areawide planning agencies to use only existing information, which in most instances was hopelessly limited. Furthermore, planning grants were frequently insufficient to enable planning agencies to retain capable consultants.

POTW PLANNING

Section 208 planning has replaced facilities planning under section 201 as the vehicle for siting, sizing, designing, and enlarging the collector sewers, interceptor sewers, sewage treatment plants, and sludge disposal facilities that we refer to as publicly-owned treatment works. Each water quality management plan must identify "treatment works necessary to meet the anticipated municipal and industrial waste-treatment needs of the area over a 25-year period." Planning agencies are responsible for establishing construction priorities and time schedules for the initiation and completion of treatment works and for instituting a regulatory program to implement the construction of treatment works and assure their adequate functioning. This program must "regulate the location, modification, and construction of any facilities within such area which may result in any discharge in such area."

In view of these requirements, one of the critical outputs of the water quality management planning process is a land-use plan encompassing time-related population and land-use projections which the planners consider compatible with water quality goals. These land-use projections should govern decisions regarding the phased development of systems and the modular construction of individual facilities to meet future needs, thus avoiding initial overdesign of waste treatment systems and consequent distortion of growth patterns. Crucial to this aspect of the regulatory program is the formation of a sewer-hookup schedule, described by federal guidelines as "important in managing the system over time in order to prevent growth from exceeding the designed capacity of the system." Each significant modification of the POTW (e.g., sewer extension or treatment plant improvement) must be approved as an amendment to the 208 plan.

Water quality management plans have provided a strong POTW planning framework, which some states are adapting to other waste treatment needs. For example, a number of states regulate all aspects of septic systems within the 208 process. Areawide agencies serve as "septic system management districts" that install septic systems, repair them, pump them out, and dispose of the septage (often at the POTW). Customers are then billed directly for these services. This model might also be appropriate for collection of "household toxics" or retrofitting of homes contaminated by naturally occurring radon gas.

Chapter 36
OCEAN DISPOSAL OF WASTE

Ocean pollution has many causes, including runoff from nonpoint sources, discharges from POTWs and industrial point sources, intentional dumping from vessels, and accidental spills. Nonpoint source control was discussed in Chapter 34, and the section 301(h) variance for POTWs making ocean discharges was discussed in Chapter 28. Cleanup of oil and hazardous substances will be analyzed in Chapter 39. This chapter will concentrate on point source ocean discharges and ocean dumping from vessels.

OCEAN DISCHARGE CRITERIA

Congress singled out point source discharges into the ocean for special consideration under the Clean Water Act. Section 403(a) of the CWA prohibits the issuance of a permit to any ocean discharger unless it is in compliance with EPA guidelines. The EPA has promulgated these guidelines as mandatory regulations.

The EPA's regulations include the seven congressional criteria for determining whether the discharge will degrade the ocean: (1) the effect on human health and welfare; (2) the effect on marine life; (3) the effect on esthetic, recreational, and economic values; (4) the persistence and permanence of the effects; (5) the effect of the disposal at varying rates, of particular volumes and concentrations; (6) other possible locations and methods of disposal or recycling, including land-based alternatives; and (7) the effect on alternate uses of the oceans, such as mineral exploitation and scientific study.

In light of these criteria, the EPA or a state issuing the permit must decide whether a discharger will cause an "unreasonable degradation of the marine environment." The term "unreasonable degradation" means (1) significant adverse effects on the ecosystem; (2) threats to human health; or (3) loss of esthetic, recreational, scientific or economic values which is unreasonable in relation to the benefit derived from the discharge. No permit may be issued if there are reasonable alternatives to ocean discharge.

If the permit issuer has insufficient information to determine that there will be no unreasonable degradation of the marine environment, he must make further analyses of effects, alternatives, and monitoring strategies. If sufficient information exists, permits must include bioassay requirements and a monitoring program "sufficient to assess the impact of the discharge on the water, sediment, and biological quality including, where appropriate, analysis of the bioaccumulative and/or persistent impact on aquatic life of the discharge."

These ocean discharge criteria are a strict water quality-based approach to regulating ocean discharges that applies over and above technology-based effluent limitations. Unlike section 303 of the CWA, no wasteload allocations are made under section 403. Instead, a discharge-by-discharge analysis is made of the potential environmental impacts. Also in contrast to section 303, reasonable alternatives to ocean discharge are considered and, if they exist, they must be adopted. This individualized analysis could become inordinately complex. Given the multiple sources of ocean pollution, and the scientific uncertainties that are intrinsic to recognizing and controlling them, how can a permit issuer ascertain whether a single discharge will unreasonably degrade the marine environment? And what is a "reasonable alternative?" This is but another example of the water quality-based approach's impracticality.

OCEAN DUMPING

The phrase "ocean dumping" involves disposal from a vessel of industrial waste, sewage sludge, or dredged spoil. It also covers the residuals from ocean incineration of wastes.

The applicable statute here is the Marine Protection, Research, and Sanctuaries Act (MPRSA), commonly known as the "Ocean Dumping Act." This statute prohibits all transportation of materials from the United States for the purpose of ocean dumping, except as authorized by a permit. It controls the dumping of any kind of matter except vessel sewage and oil (both regulated by the Clean Water Act), routine operational discharges from vessels, and placement of devices or structures in ocean waters for navigation or fisheries purposes.

The MPRSA provides for the regulation of dumping by a permit program. Power to issue permits is divided between the Environmental Protection Agency and the Army Corps of Engineers. The EPA sets criteria for evaluation of all permit applications, whatever the material to be dumped, and issues permits for the dumping of all materials except dredged spoils. The Corps, using EPA criteria, issues permits for dumping of dredged spoils. The EPA is also responsible under the MPRSA for designating and managing approved dump sites, and for limiting dumping to designated sites. "Critical areas" can be protected by the EPA's declaring them off limits to ocean dumping. The EPA can also designate "marine sanctuaries" and regulate activities within them.

The EPA Permit System

The MPRSA lists the following factors to be considered by the EPA in establishing and applying its ocean-dumping criteria: (1) the need for ocean dumping and the land disposal or recycling alternatives available; (2) the effects of ocean dumping on human health and welfare and on alternative uses of the oceans; and (3) the effects of ocean disposal on the marine environment and marine ecosystems. Under the MPRSA, dumping may be permitted if it does not "unreasonably degrade or endanger" the environment, using the above factors to determine reasonableness.

Like a number of other environmental protection statutes, the MPRSA was enacted to implement an international convention—in this case the London Dumping Convention (LDC). Accordingly, there is language in the MPRSA to the effect that no permit may be issued for dumping that is in violation of the LDC. However, the LDC does not authorize nations to balance or even consider the incremental costs and environmental impacts of land-based alternatives. In this area, the MPRSA and LDC appear to be inconsistent. The legal implications of this disjunction are as yet unclear.

EPA regulations set up various categories of wastes and types of permits. A particular class of waste can only be dumped after the permit appropriate for that class has been issued.

1. *Prohibited Materials*

These are high-level radioactive wastes, radiological, chemical, or biological warfare agents, materials of unknown components and properties, and materials which float or remain in suspension so as to cause risks to navigation, fishing, or recreation. Prohibited materials cannot be dumped under any circumstances.

2. Constituents Prohibited as Other Than Trace Contaminants

These are organohalogens, mercury, cadmium, oil not covered by the CWA, and known or suspected carcinogens, mutagens, and teratogens. "Trace amounts" are determined by listed concentrations for some wastes and bioassay results for others. If the trace contaminant levels are exceeded, these constituents can be dumped only by authority of an "emergency permit" ("unacceptable risk relating to human health and . . . no other feasible solution"), a "research permit" under certain conditions, and an "incineration at sea" permit. The EPA's proposed incineration at sea permits and regulations have been highly controversial.

3. Specially Regulated Wastes

Some of these are liquid wastes immiscible with or slightly soluble in seawater, wastes containing living organisms, highly acidic or alkaline wastes, and oxygen-demanding wastes. Restrictions on their disposal are stipulated in the regulations, but these restrictions can be exceeded in dumped materials under the terms of an emergency permit, a research permit, an incineration at sea permit, or an "interim permit." An interim permit can only be granted for otherwise nonconforming wastes if there is a need to ocean dump the material and the adverse environmental effects of the dumping are outweighed by this need or by the adverse effects of alternate disposal methods. The interim permit can only last for one year, and it can only be issued to an existing dumper. Each interim permit must include an environmental assessment and a plan for eliminating the ocean disposal or bringing the waste into compliance with EPA criteria "as soon as practicable." Industrial wastes and sewage sludge are dumped under interim permits.

4. Toxic Materials

These are defined as wastes that exceed the "limiting permissible concentration" (LPC). The LPC is

> that concentration of waste or dredged material in the receiving water which, after allowance for initial mixing . . . , will not exceed a toxicity threshold defined as 0.01 of a concentration shown to be acutely toxic to appropriate sensitive marine organisms in a bioassay carried out in accordance with approved EPA procedures.

Wastes exceeding LPC limits can be dumped under the same conditions as specially regulated wastes.

5. *Other Materials*

These can be dumped for three years under a "special permit" or indefinitely under a "general permit," unless ocean dumping is clearly unnecessary or environmentally harmful.

Evaluation of EPA Permit System

Since 1973, approximately 350 holders of interim permits have discontinued their ocean dumping. In fact, only one company is still authorized to dispose of industrial wastes by ocean dumping.

Phasing out the ocean dumping of sewage sludge, however, has been far more difficult. With so many new POTWs now on line, with a comparatively ineffectual pretreatment program, and with the EPA having yet to promulgate its sludge management regulations, sewage sludge disposal has become one of America's foremost environmental problems. Boston and Los Angeles dispose of their sludge through ocean outfall pipes, but both cities are under EPA orders to curtail this practice. New York City and a number of New York metropolitan area sewerage districts dump their sludge in the Atlantic Ocean using sludge barges.

In 1977 Congress amended the MPRSA to provide that the EPA "shall end the dumping of sewage sludge into ocean waters . . . as soon as possible but in no case may the Administrator issue any permit or renewal . . . which authorizes any such dumping after December 31, 1981." Almost everyone, including the EPA and the amendment's congressional sponsors, believed the amendment to mean that all ocean dumping of sewage sludge had to be curtailed by the end of 1981. However, in a lawsuit brought against the EPA by New York City, a federal district court ruled that the EPA could not prohibit ocean dumping of sewage sludge unless it found that sludge dumping would unreasonably degrade the marine environment, taking into consideration the need for ocean disposal, environmental impacts, and the economic and environmental costs of alternative disposal methods. Thus, sewage sludge dumping in the New York Bight continues, although the dump site has been moved offshore from 12 to 106 miles.

The Corps Permit Program

Congress has been sharply criticized for placing regulatory control in an agency which is the major producer and dumper of dredged spoils. No permit is needed for dumping spoils from dredging operations performed by or for the Corps. If a permit is necessary for another spoil dumper, the EPA can object to its issuance and the Corps can respond under a complex procedure for dispute resolution. The EPA has been sued because its criteria treat dredged wastes as inherently less harmful than nondredged wastes. There has also been extensive litigation over the EPA's choice of dumpsites for dredged and nondredged wastes.

Enforcement

The Coast Guard is responsible for conducting surveillance activities to ensure compliance with the MPRSA. It refers apparent violations to the EPA for enforcement action. In a 1977 report the General Accounting Office found serious deficiencies in Coast Guard surveillance of ocean dumpers.[27] The EPA has the authority to assess administrative fines against MPRSA violators, up to $50,000 for each offense depending on its gravity and the violator's good faith and subsequent compliance. Any person who knowingly violates the MPRSA can be held liable for a criminal fine of up to $50,000 per violation and a prison term of up to one year. Citizens may sue under the same circumstances as found in the CWA. A distinction between the MPRSA and the CWA involves the role of states. States cannot adopt or enforce any rule relating to ocean dumping. They can only propose to the EPA special criteria for dumping within state waters (three miles from shore baselines) or dumping that may affect state waters. Unlike permit responsibilities under the CWA, MPRSA regulation cannot be delegated to states.

Chapter 37
GROUNDWATER PROTECTION

There is nothing comparable for groundwater to the comprehensive federal legislation regarding discharges to surface water.

> Groundwater is a vitally important renewable resource that has been taken for granted and given little protection. Despite the fact that 45% of the nation's drinking water comes from the ground, efforts to abate and even monitor pollution of groundwater have been limited. The massive national clean-up efforts associated with the landmark environmental legislation of recent years largely ignored groundwater and, in fact, increased groundwater contamination by encouraging diversion of pollutants from the air and surface waters to the ground. . . .
> Spurred by the emerging awareness of groundwater contamination and its dangers, Congress recently enacted statutes that provide an initial framework for control of groundwater pollution, and some state and local agencies are undertaking their own protection efforts. Unfortunately, the federal framework is a confusing hodgepodge and the state and local ventures are far from comprehensive.[28]

Controlling groundwater pollution will be the first priority of future water pollution control law.

This chapter will begin with a discussion of the EPA's "Ground-Water Protection Strategy." Federal and state efforts to protect groundwater will then be scrutinized.

EPA GROUNDWATER PROTECTION STRATEGY

In 1984 the EPA issued its final "Ground-Water Protection Strategy" (Strategy). The Strategy is based on the finding that three levels of government—local, state, and federal—have substantial responsibility for groundwater protection, but that state and local governments should have the principal role in groundwater protection and management.

According to the Strategy, the EPA's role would be to (1) foster stronger state government programs for groundwater protection, (2) cope with inadequately addressed problems of groundwater contamination, (3) establish a framework for decisionmaking in EPA programs, and (4) strengthen its own internal groundwater organization.

Strengthening state programs involves (1) encouraging states to make use of existing grant programs—especially Clean Water Act program grants—to develop groundwater protection programs and strategies; (2) providing states with technical assistance in solving specific groundwater problems; and (3) supporting a strong research program in groundwater. One of the inadequately addressed groundwater problems that the EPA has been studying is contamination by pesticides and nitrates from agriculture.

Next to its philosophy of state and local primacy in the groundwater protection area, the Strategy's most important element is its policy framework for guiding EPA programs. EPA policy will be "based on the recognition of the highest beneficial use to which the ground-water resource can presently or potentially be put." Aquifers will be divided into three classes: (1) special groundwaters, (2) current and potential sources of drinking water and water having other beneficial uses, and (3) groundwater not a potential source of drinking water and of limited beneficial use. Special groundwaters are irreplaceable sources of drinking water and ecologically vital waters. These aquifers will be protected by the EPA under an "antidegradation" standard. Most of the usable groundwater in the United States will fall into the second class. For aquifers in this class, "prevention of contamination will generally be provided through application of design and operating requirements based on technology, rather than through restrictions on siting. . . ." The third class of aquifers includes groundwaters with a TDS level over 10,000 mgL and groundwaters that are so contaminated by naturally occurring contaminants and by human activity that they cannot reasonably be made potable. These aquifers will receive less protection as long as they are unconnected to aquifers in the higher two classes.

In order to carry out its groundwater protection activities most efficiently, the EPA has established an Office of Groundwater Protection in the Office of Water and counterpart offices in each region.

The EPA's Strategy has been criticized on a number of grounds. First, groundwater pollution is often an interstate problem:

Polluted groundwater migrates across state boundaries. It may affect travel and commerce by entering surface waters; or it may affect travellers, who have to drink it. Conflicting standards for groundwater quality may hamper interstate business or may incite economically and socially disruptive industry migrations. They may also make the health of our nation's citizens dependent on where they live. Simultaneously, the proliferation of varying local rules may impede the development of new technologies and practices which would be in the federal interest.[29]

Will the Strategy do anything to prevent the "tragedy of the commons" that is currently being played out above the Ogallala aquifer?

Second, there is doubt about whether the EPA can provide states with the research, technical, and financial support to encourage them to take a more active role in protecting groundwater quality.

Third, critics believe that the EPA's "beneficial use" policy is inadequate. It assumes that future drinking water needs and treatment capabilities can be reliably predicted:

Of course, more than ordinary 'judgment' will be required to accurately forecast patterns of growth, development and resettlement, competing uses, or the availability of alternate supplies, say, fifty or one hundred years into the future. The plain truth is that current use designations often will determine future uses, since it may be difficult to adjust the water quality in a given aquifer to accommodate changing needs.[29]

Professor Dycus, among others, calls for management of all aquifers under an antidegradation standard.

FEDERAL STATUTORY PROGRAMS

The Office of Technology Assessment has categorized sources of groundwater contamination as follows: (1) sources designed to discharge substances; (2) unplanned releases from sources designed to store, treat, and/or dispose of substances; (3) sources designed to retain substances during transport or transmission; (4) sources discharging substances as a consequence of other planned activities; (5) sources providing conduit or inducing discharge through altered flow patterns; and (6) naturally occurring sources whose discharge is created and/or exacerbated by human activity.[30]

Sources designed to discharge substances include subsurface percolation (e.g., septic systems and cesspools), injection wells, and land application of wastewater and sludge. Subsurface percolation is not regulated at the federal level. Land application of wastewater and sludge is regulated under the Clean Water Act. Injection wells are regulated through the Underground Injection Control (UIC) Program authorized by the Safe Drinking

Water Act. "Underground injection" is defined as underground disposal of liquid wastes by discharge into wells that are deeper than they are wide. Injection of brine or other fluids from oil and gas operations and underground injection for secondary or tertiary recovery of hydrocarbons can only be regulated where "essential" to protect drinking water.

The definition of "underground source of drinking water" (USDW) is central to the UIC program. An USDW is an aquifer that (1) supplies or could supply a public water system, (2) currently supplies drinking water for human consumption, or (3) contains fewer than 10,000 milligrams per liter total dissolved solids. However, an aquifer that meets one of these criteria may be exempted from the program if it does not presently serve as a source of drinking water and could not because (1) it is actually or potentially mineral, hydrocarbon, or geothermal energy producing, (2) it is situated at a depth or location that makes recovery of water for drinking water purposes economically or technologically impractical, (3) it is untreatably contaminated, (4) it is located over a mining area subject to subsidence or collapse, or (5) the total dissolved solids content is more than 3,000 and less than 10,000 milligrams per liter and the groundwater is not reasonably expected to supply a public water system.

EPA regulations establish five classes of injection wells, based on purpose and degree of threat to an USDW. Each class of injection well is subject to requirements for construction, operation, monitoring and reporting, plugging and abandonment, financial responsibility, and mechanical integrity. All injection wells must be authorized by permit or regulation. If a state does not have an approved UIC program, the EPA will carry out a program for the state. Wells that inject hazardous waste directly into, above, or near an USDW (i.e., within one-fourth mile) are prohibited. Each permit applicant bears the burden of showing that his discharge will not endanger an USDW. Under the 1984 amendments to the Resource Conservation and Recovery Act (RCRA), deep well injection of certain hazardous wastes is prohibited entirely.

In addition to the UIC program, the Safe Drinking Water Act includes three other groundwater protection programs.

The EPA, either by petition or on its own initiative, may designate an aquifer as the "sole or principal source of drinking water" for an area. After designation, "no commitment for federal financial assistance may be entered into for any project which [the EPA] determines may contaminate such aquifer through a recharge zone so as to create a significant hazard to public health." Projects that typically receive federal assistance are highways, housing developments, and sewage treatment plants. Approximately 20 "sole source aquifers" have been designated. However, the program does not apply to federal construction projects such as water resource development projects. In addition, the designated area might not include an upstream headwater area that provides recharge to the aquifer.[31]

Under the Critical Aquifer Demonstration Program, the EPA has promulgated regulations establishing criteria for identifying critical aquifer protection areas: vulnerability to contamination; use as a drinking water source; and economic, social, and environmental factors are to be considered. Each area must lie within an area that has been designated as a sole or principal source aquifer. Any governmental entity or planning entity (including an interstate planning entity) having jurisdiction over a critical aquifer protection area may apply to the EPA for a demonstration grant. State governors must be joint applicants. A grant application must contain area boundaries, procedures for public participation, an assessment of surface and ground water resources within the area, a comprehensive management plan, and a set of implementation measures and schedules. The comprehensive plan is intended to "maintain the quality of the ground water in the critical protection area in a manner reasonably expected to protect human health, the environment and ground water resources." Central to the comprehensive plan are an assessment of the relationship between land uses and ground water quality, identification of management practices to protect ground water quality, and specification of implementation authorities, program costs, and potential funding sources. If the EPA approves the application, a matching grant may be made of up to 50% of the costs—not to exceed $4 million in any fiscal year—of developing and implementing the plan. Total authorized matching grant funds are $10 million for fiscal 1987, $15 million for fiscal 1988, and $17.5 million for each of 1989, 1990, and 1991.

State programs to protect wellhead areas within their jurisdictions from contaminants that may have any adverse effect on the health of persons are now eligible for 50% to 90% federal program grants. The term "wellhead protection area" means the "surface and subsurface area surrounding a water well or wellfield, supplying a public water system, through which contaminants are reasonably likely to move toward and reach such water well or wellfield." States are to identify wellhead protection areas based on EPA technical guidelines. State programs must also (1) identify potential pollution sources within each wellhead protection area; (2) describe a program that contains, as appropriate, technical assistance, financial assistance, control measures, education, training, and demonstration projects; (3) specify duties of governmental agencies and water purveyors; (4) include contingency plans for alternate drinking water supplies for each public water system in the event of well contamination; and (5) establish a procedure for considering potential contamination in reviewing new public water system well permits. If the EPA does not disapprove a state program within nine months of submission, it is deemed approved. Each state must make "every reasonable effort" to implement the program within two years after approval. Congress has authorized for matching grants to states for wellhead area protection a total of $20 million for each of fiscal years 1987 and 1988, and $35 million for each of 1989, 1990, and 1991.

Sources designed to store, treat, and/or dispose of substances (discharge through unplanned releases) include landfills, open dumps, surface impoundments, waste tailings and piles, materials stockpiles, human and animal graveyards, aboveground and underground storage tanks, containers, open burning, radioactive disposal sites, and residential hazardous waste disposal.

Under the Uranium Mill Tailings Regulation Control Act, the Nuclear Regulatory Commission is empowered to control leaching from active uranium mill tailings piles and the U.S. Department of Energy is responsibile for cleaning up tailings left at abandoned sites. Industrial material stockpiles can be regulated by including Best Management Practices for "ancillary pollution" in discharge permits of industrial point sources. The EPA is responsible, under the Nuclear Waste Policy Act, for promulgating standards protecting the general public from offsite releases from high-level radioactive material in repositories. The Toxic Substances Control Act regulates all discharges of PCBs, and might also be used to regulate residential disposal of toxic wastes.

By far the most important federal statute in this area, however, is the Resource Conservation and Recovery Act (RCRA). The RCRA extends federal regulation to land disposal of municipal waste and generation, transportation, treatment, storage, and disposal of hazardous waste. The RCRA's definition of "disposal" specifically refers to groundwater. However, the RCRA does not apply to storage of hazardous raw materials or products.

With regard to municipal solid waste disposal, the RCRA requires the EPA to promulgate regulations distinguishing sanitary landfills from "open dumps." New open dumps are prohibited and existing dumps must be closed or upgraded. Sanitary landfills located above USDWs must not endanger the quality of groundwater.

The RCRA's approach to hazardous waste management consists of four major elements: (1) federal identification of hazardous wastes; (2) a manifest system for tracing hazardous wastes from generator, to transporter, to treatment, storage, or disposal facility; (3) federal minimum standards for hazardous waste treatment, storage, and disposal, enforced through a permit system; and (4) state implementation of hazardous waste management programs at least equivalent to the federal program.

Hazardous wastes are listed by chemical constituent and by waste stream. But if a waste does not appear on these lists it may still be hazardous, and it is the responsibility of each generator to test its waste according to EPA regulations. One such test simulates leaching into groundwater. Generators of hazardous wastes must also (1) keep records and report to the federal government, (2) initiate a manifest system "to assure that all such hazardous waste generated is designated for treatment, storage, or disposal in . . . facilities . . . for which a permit has been issued," and (3) properly label and containerize hazardous wastes delivered to transporters and treatment, storage, and disposal facilities. The duties of a transporter involve (1) record keeping and reporting, (2) accepting only properly labeled and containerized wastes, (3) complying with the manifest system, and most important, (4) transporting all hazardous waste only to the permitted facility that the generator identifies on the manifest. The manifest system terminates with the receipt of the wastes by the owner or operator of the disposal facility and his notification to the generator.

Permits are not required of generators or transporters of hazardous wastes, but disposal, treatment, or storage of these wastes are prohibited except in accordance with a permit. In order to obtain and retain a permit, an owner or operator of a disposal, treatment, or storage facility must meet EPA performance standards governing location, design, construction, operation, and maintenance of such a facility. The applicable performance standards become permit conditions, in addition to the record keeping and reporting requirements.

A state is authorized to administer and enforce a hazardous waste regulatory program in lieu of the federal program if the EPA finds that the state program is equivalent to the federal program and is consistent with the programs of neighboring states. No state may impose less stringent requirements than those included in federal law, but a state may elect to be more stringent. The EPA is the primary enforcement authority where a state program has not been approved. Even subsequent to approval, the EPA can enforce directly against a violator after giving notice to the state. If an approved program is later determined to be inadequate, the EPA may withdraw program authorization and reinstitute the federal program in that state.

The RCRA provides the EPA with a broad range of enforcement mechanisms, including compliance orders, administrative fines, civil actions for injunctions, civil penalties of up to $25,000 per day of violation, permit suspension or revocation, and criminal indictments for knowing transportation to an unpermitted facility, unpermitted disposal, and making false statements in applications, manifests, labels, and reports. State programs must ensure adequate enforcement, but state enforcement tools may be weaker than those which the RCRA provides the EPA.

The EPA has promulgated regulations applicable to facilities that treat, store, or dispose of hazardous wastes. These regulations cover facility performance standards, preparedness and prevention, contingency plans and emergency procedures, the manifest system, closure and postclosure measures, financial responsibility, and the management of containers, tanks, surface impoundments, waste piles, land treatment units, landfills, incinerators, thermal treatment units, and chemical, physical, and biological treatment units. An entire subpart of these regulations is devoted to groundwater protection. The facility permit must contain conditions to ensure that hazardous constituents from a regulated unit do not exceed certain concentration limits at the downgradient facility boundary for the active life of the facility and during postclosure if the site has not been decontaminated on closure. A "hazardous constituent" is a constituent on the RCRA list which has been detected in the underlying aquifer or "that [is] reasonably expected to be in or derived from waste contained in a regulated unit." However, the EPA may grant a variance if it "finds that the constituent is not capable of posing a substantial present or potential hazard to human health or the environment." "Concentration limits" are background levels, drinking water standards where they exist, or alternate concentration limits based on the variance criterion for a hazardous constituent. In other words, permit limitations are set on a site-specific basis. Moreover, a site is exempt from regulation if the EPA finds "that there is no potential for migration of liquid from a regulated unit to the uppermost aquifer during the active life . . . and the post-closure care period."

All facilities treating, storing, or disposing of hazardous wastes must perform detection monitoring. The EPA may exempt from groundwater monitoring requirements land disposal units that are both designed to prevent liquids from entering the unit and are equipped with multiple leak detection systems. Any waste found beyond the actual waste management area is presumed to originate from the regulated unit. When a hazardous constituent is detected beyond the facility boundary, the owner or operator must take "corrective action," even beyond the boundary, to remove the waste constituents or treat them in place. A compliance monitoring system is also required.

Amendments to the RCRA enacted in 1984 significantly strengthen its protection of groundwater. First, the land disposal of a hazardous waste, including deep well injection, must be banned unless the EPA determines that a particular method of land disposal will be consistent with protecting human health and the environment. A method of land disposal may not be determined to be protective of human health and the environment unless a petitioner demonstrates that there will be no migration from the land disposal unit for as long as the waste remains hazardous. A disposability determination must be made within 66 months, or land disposal of the hazardous waste is automatically banned. Variances for up to two years may be obtained. Second, the landfilling of noncontainerized liquids is prohibited, and landfilling of containerized liquids must be minimized. Third, facilities permitted under "interim status" must either install double liners, leachate collection systems, and groundwater monitoring devices or stop receiving, storing, or treating hazardous waste. Numerous exceptions and modifications are available. However, impoundments with clay liners must close as storage impoundments. And finally, a new facility must have a double liner with leachate collection above (for landfills) and between the liners, and must monitor groundwater. A variance from the double liner requirement is available where owners and operators can show that an alternate design, together with locational characteristics, is as effective in preventing migration of hazardous constituents to groundwater.

The 1984 amendments also establish a regulatory program to control underground storage tanks containing hazardous substances and petroleum products. Excluded, among other things, are farm and residential tanks storing motor fuels, heating oil tanks used for noncommercial purposes, septic tanks, regulated pipelines, surface impoundments, stormwater and wastewater collection systems, and flow-through process tanks. The regulatory program for existing tanks includes notification by tank owners and operators, performance standards, leak detection or inventory control system and tank testing, recordkeeping and reporting, corrective action, financial responsibility, and closure requirements. For new tanks, the regulations include requirements for design, construction, installation, release detection, and compatibility standards. States may be delegated authority to administer this program.

Sources designed to retain substances during transport or transmission include pipelines and materials transport and transfer operations. Under the Hazardous Liquid Pipeline Safety Act, the Department of Transportation (DOT) promulgates regulations for the movement of hazardous liquids by pipeline and their storage incidental to such movement. DOT also administers the Hazardous Material Transportation Act, under which standards are set for transporting hazardous materials.

Sources discharging substances as consequences of other planned activities include irrigation practices, pesticide and fertilizer applications, animal feedlots, de-icing salts applications, urban runoff, percolation of atmospheric pollutants, and mining and mine drainage. Of these, only groundwater contamination by surface mining and application of agricultural pesticides are regulated by the federal government.

Under the Surface Mining Control and Reclamation Act (SMCRA), administered by the Secretary of the Interior, groundwater in surface-mined and adjoining areas must be retained in, or restored to, its original condition after mining. Surface mine operators must either drill new wells or provide alternate water supplies during mining. Before and during mining, monitoring of groundwater levels, infiltration rates, subsurface flow, storage characteristics, and groundwater quality is required. If surface mining will irretrievably damage an aquifer, the area may be designated as unsuitable for mining. Before receiving a mining permit from the Interior Department or a state administering the program, an operator must file a detailed mining plan including hydrologic impacts, mitigation actions, and monitoring programs. A reclamation bond must also be posted.

Although it does not directly regulate groundwater pollution by agricultural pesticides, the Federal Insecticide, Fungicide, and Rodenticide Act might be used to ban a pesticide that contaminates groundwater. Alternatively, the pesticide might be registered subject to labeling or application requirements.

Sources providing conduit or inducing discharge through altered flow patterns include production wells, monitoring and exploration wells, and construction excavation. None of these is federally regulated. Neither are groundwater-surface water interactions, natural leaching, and salt-water intrusion—which are categorized by the OTA as *naturally occurring sources whose discharge is created and/or exacerbated by human activity.*

The federal statutes that regulate groundwater discharges do not constitute a comprehensive or consistent federal groundwater protection program. First, many activities that pollute groundwater are not regulated by the federal government. Second, the federal statutes discussed above are administered by a number of federal agencies with different statutory missions and levels of environmental sensitivity. Even those programs administered by the EPA cannot always be integrated.

The EPA's reaction to the fragmented federal groundwater protection situation has been to integrate as well as possible and leave the rest to the states.

STATE GROUNDWATER PROTECTION PROGRAMS

The Environmental Protection Agency has identified eight categories of state groundwater protection legislation:[32]

Statewide strategies require a state agency or group of state agencies to develop a comprehensive plan to protect groundwater quality from contaminant sources in any area of the state. The strategy may apply existing regulatory authority to groundwater problems or recommend the enactment of new legislation.

Aquifer classification involves identifying and classifying aquifers—either based on antidegradation or projected uses—to determine the degree of protection that is necessary. Classification can be used as a planning tool to set groundwater quality standards, discharge permit requirements, and land-use controls. At least 22 states have adopted classification systems. Some states, Maine and New Hampshire among them, impose an antidegradation requirement on all groundwater.[33]

Groundwater Quality Standard-Setting establishes threshold levels for aquifer contamination. Groundwater quality standards are ambient standards based on either natural background or projected uses. Most standards are numerical, but some are narrative. Groundwater quality standards can be applied uniformly across a state or differentially in specific areas within the state.[33]

Source controls are imposed by most states. They consist of "performance standards" (specifying how much pollution a source can release), "technical standards" (specifying particular technologies that must be adopted), or best management practices. "In groundwater protection programs performance standards may apply directly to the facility's discharge, or to the amount of contamination allowed to leave a *zone of discharge* allowing for the diffusion and attenuation of wastes within a strictly limited area underlying a waste-producing facility."[33] Some states require discharge permits, incorporating source standards, for all dischargers to groundwater. States impose source controls when they implement federal programs such as UIC or RCRA, but some go beyond federal law in regulating landfills, hazardous waste disposal, septic systems (usually by delegation to local governments), agricultural pesticide use, and disposal of residential hazardous wastes.

Land-Use Controls are traditionally imposed by local governments. They can be used to prevent development in aquifer recharge areas that might lead to contamination of aquifers. Some of these strategies are hydrologic zoning, performance standard zoning, acquisition of land or conservation easements, and transfers of development rights. Land-use controls to protect groundwater quality are rare in the United States.

Groundwater Funds, as distinguished from state "superfunds," are specific financial accounts established to protect groundwater quality. Revenue in these funds may be used for groundwater cleanup, replacement of drinking water supplies, or groundwater monitoring. For example, an Iowa statute creates a groundwater fund comprised of solid waste tonnage fees to be used for monitoring groundwater quality and developing groundwater quality standards.

Underground storage tank controls have been implemented by many states in advance of the federal program established by the 1984 RCRA amendments.

Water Use Management incorporates groundwater quality protection in the criteria used to justify more stringent water allocation measures in designated critical areas, such as those where salt water is intruding into potable water aquifers. This kind of comprehensive management is also comparatively rare (see Chapter 5).

Like federal programs to protect groundwater quality, most state programs are limited and sporadic. Some of the potentially groundwater-contaminating activities that the majority of states do not adequately regulate are agricalutural pesticide and fertilizer use, development in freshwater wetlands, urban stormwater runoff, construction-site runoff, septic systems, residential hazardous waste disposal, nonhazardous waste landfills, irrigation practices, animal feedlots, raw materials stockpiles, road de-icing, deep mining, production wells, and—perhaps most important—development in aquifer recharge areas. Moreover, state implementation of federal groundwater protection programs is frequently deficient.

Only time will tell whether the current system—or lack of one—can protect America's groundwater.

Chapter 38

OIL AND HAZARDOUS SUBSTANCES CLEANUP

Prevention is, of course, the most efficient approach to controlling water pollution. But when a pollution incident has occurred, it may be necessary to clean up the area in order to protect the public or the ecosystem. Predictable cleanup liability is also a deterrent to pollution:

> Because of the perceived inadequacies of control strategies relying exclusively on regulation, a number of alternative international, federal, and state mechanisms have been developed which emphasize liability as an economic disincentive to pollution. These approaches foster the internalization of "social" costs of pollution by those responsible for oil and hazardous substances releases. By incorporating the "polluter pays" principle, such legal mechanisms are intended both to deter harmful conduct and to redress pollution-caused harm.[34]

This chapter primarily examines federal law dealing with oil and hazardous substances cleanup. The active international law in this field is beyond the scope of this book.

OIL AND HAZARDOUS SUBSTANCES SPILLS

Liability for oil spills is covered under four federal statutes. Section 311 of the CWA is the most comprehensive of these provisions. The others were enacted to control oil pollution in specific areas. They are the Outer Continental Shelf Lands Act Amendments, the Deepwater Port Act, and the Trans-Alaska Pipeline Authorization Act. These four provisions are similar in that they establish trust funds to pay for cleanup of oil spills. However, they differ in many ways, and Congress has been considering a comprehensive oil pollution control statute.

The coverage of section 311 is limited to "spilling, leaking, pumping, pouring, emitting, emptying, or dumping" oil and hazardous substances from vessels, onshore facilities, and offshore facilities into fresh or marine navigable waters of the United States seaward to the 200-mile limit. It does not cover groundwater discharges or point source discharges under CWA discharge permits. Section 311 is administered by the Environmental Protection Agency and the Coast Guard.

The first step in the 311 process is for the EPA to determine "those quantities of oil and any hazardous substances the discharge of which may be harmful to the public health or welfare." In 1979, the EPA promulgated regulations designating approximately 300 substances as hazardous and establishing a "reportable quantity" for each. Virtually all discharges of oil are reportable. Any person in charge of a vessel or an onshore or offshore facility is required to "immediately notify" the Coast Guard or EPA "as soon as he has knowledge" of the discharge of a reportable quantity of oil or a hazardous substance. Failure to notify is punishable by fines and imprisonment.

Apart from the notification requirement, discharge of a reportable quantity gives rise to responsibility for both penalties and the costs of removal. There is an alternative set of procedures to assess civil penalties for hazardous waste spills: an administrative assessment by the Coast Guard of up to $5,000 or a civil lawsuit by the EPA to assess a penalty of up to $50,000 for each discharge of a reportable quantity. Where the EPA can show that the discharge was the result of "willful negligence or willful misconduct within the privity and knowledge" of the responsible person, the maximum penalty is $250,000. Lesser penalties apply to oil spills.

As for cleanup liability, whenever any oil or a hazardous substance is discharged the Coast Guard and EPA are authorized by section 311 "to remove or arrange for the removal of such oil or substances at any time" unless they determine that "such removal will be done properly by the owner or operator" of the vessel or onshore or offshore facility from which the discharge occurs. In addition to removal, the federal government "may act to mitigate the damage to the public health or welfare caused by such discharge." An owner or operator is liable to reimburse the federal government "for the actual costs" incurred by federal officials or their contractors for the removal of oil or a hazardous substance that is discharged in reportable quantities, including the costs of mitigation. Limitations on liability differ among onshore facilities, offshore facilities, and vessels and between dischargers of oil and hazardous substances. However, an owner or operator is liable for the full amount of the costs where the United States can show that the discharge "was the result of willful negligence or willful misconduct within the privity and knowledge" of the responsible person. Removal costs include restoration or replacement of natural resources damaged or destroyed by the discharge. Under section 311, a discharger is exempt from cleanup liability only if he can prove that the discharge "was caused solely by (a) an act of God, (b) an act of war, (c) negligence on the part of the United States Government, or (d) an act or omission of a third party without regard to whether such act or omission was or was not negligent, or any combination of the foregoing causes." This is often referred to as "strict liability" because the exercise of care is no defense.

In order to prevent oil spills before they occur, owners and operators of large oil storage facilities must develop, implement, and maintain spill prevention, control, and countermeasure (SPCC) plans. Under CERCLA, emergency response planning for facilities that utilize extremely hazardous substances is performed by local emergency planning committees. Vessels are encouraged to take preventative measures by requiring them to establish proof of financial responsibility to meet potential liability under section 311.

Section 311 includes a revolving fund, financed by fine receipts and congressional appropriations, to support federal oil spill cleanup efforts where financially responsible dischargers that are willing to undertake cleanup cannot be located. The 311 fund may not be used to compensate "third-party" spill victims, that is, owners of shorefront property, fishermen, and others damaged by spills.

CERCLA AND HAZARDOUS WASTES

The Comprehensive Environmental Response, Compensation, and Liability Act (CERCLA), also known as "Superfund," applies to "releases" of "hazardous substances" from "facilities" and vessels. The term "release" is extraordinarily comprehensive, covering "any spilling, leaking, pumping, pouring, emitting, emptying, discharging, injecting, escaping, leaching, dumping, or depositing into the environment." Abandoned hazardous waste containers are also releases. Unlike section 311, the CERCLA applies to groundwater discharges. "Release" does not include specified discharges of nuclear materials, workplace emissions, most engine exhausts, and "normal" fertilizer applications. Moreover, the releases of pollutants under federal permits, releases from the applications of registered pesticides, and releases mentioned in environmental impact statements are exempt from the CERCLA.

"Hazardous substances" are any substances listed as toxic or hazardous under any federal pollution control statute. Oil and gas are not "hazardous substances." A "facility" is:

(A) any building, structure, installation, equipment, pipe or pipeline (including any pipe into a sewer or publicly owned treatment works), well, pit, pond, lagoon, impoundment, ditch, landfill, storage container, motor vehicle, rolling stock, or aircraft, or (B) any site or area where a hazardous substance has been deposited, stored, disposed of, or placed, or otherwise come to be located; but does not include any consumer product in consumer use or any vessel.

Abandoned hazardous waste sites are covered by this definition.

The CERCLA contains spill notification provisions that are similar to those of section 311. Moreover, owners and operators of vessels or facilities handling hazardous wastes must show proof of financial responsibility. Unless the Coast Guard and EPA determine that the person responsible for the spill will clean it up, the federal agencies may arrange for pollution removal and remedial operations whenever "any hazardous substance is released or there is a substantial threat of such a release into the environment."

The CERCLA establishes an $8.5 billion fund over five years to finance Superfund response activities. The sources of revenue for the fund are (1) a petroleum tax of 8.2 cents/barrel for domestic crude oil and 11.7 cents/barrel for imported petroleum products (including imported crude oil)—expected to raise $2.75 billion; (2) a chemical feedstock tax—expected to raise $1.4 billion; (3) an environmental tax based on corporate minimum taxable income, at a rate of 0.12% over $2 million—expected to raise $2.5 billion; (4) general revenues of $1.25 billion; and (5) fund interest and cost recoveries—expected to raise $600 million.

Where a release has occurred, the EPA can use fund monies to clean up the site and then proceed against responsible parties for reimbursement, or move against responsible parties in the first instance. The CERCLA imposes strict, "joint and several" (recovery may be obtained from one or all) liability against (1) current owners or operators of facilities, (2) owners or operators at the time the hazardous substances were released, and (3) generators and transporters of the hazardous substances that were ultimately released by the facility. Only innocent purchasers who have made reasonable investigations are insulated from cleanup liability. The CERCLA contains dollar limitations on liability, but these ceilings are considerably higher than those under section 311.

Responsible parties may also be required to pay for "restoring, replacing, rehabilitating, or acquiring the substantial equivalent of damaged natural resources," but fund monies cannot be used for this purpose. Only natural resources belonging to, managed by, or protected by a state or the federal government are eligible. Moreover, there is no cleanup liability for resource damage occurring "wholly before" the CERCLA's enactment, or for losses arising from "long-term exposure to ambient concentrations of air pollutants from multiple or diffuse sources" (e.g., acid precipitation). Like section 311, the CERCLA does not provide for compensating private victims of hazardous waste releases.

Federal hazardous substances response must, "to the greatest extent possible," be consistent with the National Contingency Plan (NCP) begun under section 311 and updated under the CERCLA. The NCP includes preferable removal methods, cost-effectiveness criteria, the roles of governmental units, and criteria for determining cleanup priorities. Perhaps most important, the NCP shall "list . . . national priorities among the known releases or threatened releases throughout the United States," subject to annual revision. Short-term removal actions may be financed exclusively from the fund, but states must bear at least 10% of the cleanup costs and all future site maintenance expenses where long-term remedial activities are concerned. Fund monies may not be used for remedial actions where a state does not have adequate facilities for disposing of all its hazardous wastes for 20 years.

The CERCLA amendments enacted in 1986 have a significant bearing on how quickly, and to what extent, National Priority List (NPL) sites will be cleaned up. Remedial investigation/feasibility studies for facilities on the NPL must be commenced at the following rate: 275 within three years; an additional 175 within four years; and an additional 200 within five years; for a total of 650 by 1991. Substantial and continuous physical, on-site, new remedial actions at facilities on the NPL must be commenced at the rate of 175 during the first three years after enactment and an additional 200 during the following two years. There are no time limits on completion of studies or remedial actions.

The amendments require the EPA to select, to the maximum extent practicable, remedial actions that utilize permanent solutions and alternative treatment technologies or resource recovery technologies. They establish a preference for remedial actions that utilize treatment to permanently and significantly reduce the volume, toxicity, or mobility of hazardous substances. Offsite transport and disposal without treatment is the least preferred option where practicable treatment technologies are available. If the selected remedy does not achieve the preference for treatment, the EPA must publish an explanation.

For onsite actions, the amendments require remedial actions to at least attain legally applicable or relevant and appropriate federal and state standards, requirements, criteria, or limitations (ARARs), unless such requirements are waived. Maximum contaminant level goals under the Safe Drinking Water Act and water quality standards under the Clean Water Act must be met where relevant and appropriate. Alternate concentration limits are prohibited if the process assumes a point of human exposure beyond the boundaries of the facility. More stringent, promulgated state requirements must also be met. ARARs may be waived where (1) the remedial action is part of a larger one that will comply, (2) compliance will result in greater risk to human health and the environment than other options, (3) compliance is technically impracticable from an engineering standpoint, (4) an alternate remedial action will attain the equivalent of the ARAR, (5) a state has not applied its requirement consistently, or (6) a remedy would be so costly as to drain the fund of money needed to protect public health at other sites.

The amendments allow the transfer of hazardous substances offsite only to a nonleaking facility operating in compliance with RCRA and all applicable state requirements.

As an adjunct to the regulation of underground storage tanks under RCRA, the 1986 CERCLA amendments establish an Underground Storage Tank Trust Fund financed by a tax of 0.1 cents per gallon on petroleum fuels. The EPA or a state may require an owner or operator of a tank to undertake corrective action if the action will be done properly and promptly, or the agency may undertake the corrective action if necessary to protect human health and the environment. Trust fund monies may be used for financing corrective action, enforcement, or cost recovery proceedings. Priorities for action shall be those instances where the EPA or a state cannot identify a solvent owner/operator who will properly undertake the action, and where the greatest threats are posed to human health and the environment. After maximum contaminant levels are promulgated under the Safe Drinking Water Act, actions or orders must comply with them. Corrective actions may include temporary or permanent relocation of residents and installation of alternative household water supplies.

STATE SPILL STATUTES

State spill compensation and control statutes are important because (1) they impose liability where the CERCLA does not (for example, for third-party damage), and (2) they provide extra money to finance the state share of CERCLA cleanup, or to clean up sites which are not listed on the NPL. An estimated 23 states have statutory funding mechanisms to cover cleanup costs. Moreover, 13 states have reportedly enacted statutes allowing private citizens to recover for damages caused by hazardous substance spills.[34]

The New Jersey Spill Compensation and Control Act, for example, was enacted in 1976, four years before the CERCLA. The act establishes a revolving $50 million New Jersey Spill Compensation Fund, financed by a tax on major facilities that "refine, produce, store, handle, transfer, process, or transport" oil or hazardous substances. The fund may be used to remove actual discharges or threatened discharges if hazardous materials are dangerously stored or transported. Discharges that occurred prior to the act's passage may also be cleaned up, but there are limits on how much may be spent to clean up preenactment sites in any one year.

The fund is "strictly liable, without regard to fault, for all cleanup and removal costs and for all direct and indirect damages, no matter by whom sustained." Compensable damages include (1) costs of restoring and replacing natural resources, (2) costs of restoring, repairing, or replacing private property, (3) loss of income due to property or resource damage, (4) loss of tax revenues by state or local governments for one year due to property damage, and (5) interest on loans taken out by a claimant pending fund payments. The act does not mention personal injury costs, but would probably be construed to cover them.

"Any person who has discharged a hazardous substance or is in any way responsible for any hazardous substance which the department has removed" is strictly liable for all cleanup and removal costs. Where damage costs are concerned, there are dollar liability limitations similar to those in section 311. The only defenses to liability are acts of war, sabotage, and acts of God. The act has been held to apply retroactively, that is, to impose liability for preenactment discharges that are still a threat to public health.

Chapter 39
DRINKING WATER PROTECTION

Ambient water quality and tap water quality are regulated by different federal statutes. The Clean Water Act has as its goal fishable/swimmable, not potable, waters. Nevertheless, state antidegradation policies are intended to maintain the quality of some waters that are of potable quality or could be after treatment. The underground injection control program and the Resource Conservation and Recovery Act have been used to protect some aquifers that do or could serve primarily as drinking water sources. But these statutes cannot assure that tap water is fit to drink. Background contamination, incomplete regulation, nonpoint sources, and permit violations make it necessary to create another layer of protection for the public. This is the function of the Safe Drinking Water Act (SDWA) and, to a lesser extent, the Saline Water Conversion Program.

SAFE DRINKING WATER ACT

The SDWA was enacted in 1974 in response to the discovery of toxic substances in the drinking water supplies of major American cities. The Environmental Protection Agency is required to promulgate national primary and secondary drinking water regulations applicable to "public water systems," defined as systems that have at least 15 service connections or regularly serve at least 25 individuals at least 60 days per year. "Primary drinking water regulations" identify potential toxic contaminants and for each contaminant set a maximum contaminant level (MCL) if the contaminant can feasibly be measured or a treatment technique if it cannot. Secondary drinking water standards set MCLs for nontoxic contaminants that affect, for example, color and taste. As with the CWA or RCRA, states may assume primary enforcement responsibility under the SDWA and receive federal program grants.

Under the SDWA as originally enacted, the EPA was to set primary drinking water regulations in three steps. In the first step, the National Interim Primary Drinking Water Regulations became effective in 1977. (They were made final in 1986.) The EPA based the MCLs contained in these interim regulations in large part on the 1962 U.S. Public Health Service drinking water standards. In the second step, the EPA was to develop and promulgate recommended maximum contaminant levels, now known as "maximum contaminant level goals" (MCLGs). MCLGs are nonenforceable health goals for public water systems. They are to be set at a level at which, in the Administrator's judgment, "no known or anticipated adverse effects on the health of persons occur and which allows an adequate margin of safety." Finally, after MCLGs had been promulgated, the EPA was to promulgate enforceable primary drinking water regulations including MCLs and monitoring and reporting requirements for those contaminants that might have an adverse effect on human health. MCLs in the primary regulations must be set as close to MCLGs as feasible. "Feasible" means "with the use of the best technology, treatment techniques, and other means, which the Administrator finds are generally available (taking costs into consideration)." The primary regulations must then be reviewed and, if feasible, strengthened every three years.

The EPA made virtually no progress in promulgating MCLGs and MCLs until 1985, when Congress begin voicing its displeasure with the EPA's attenuated pace. On November 13, 1985, the EPA promulgated MCLGs and proposed MCLs for eight volatile synthetic organic chemicals (VOCs). MCLGs for five of these VOCs, which are suspected of being human carcinogens, were set at zero. MCLGs are critical because they are used not only in setting drinking water standards but also as guides in determining the adequacy of CERCLA cleanups. Consequently, industry challenged the zero MCLGs for the five VOCs in court, but the MCLGs were upheld by the U.S. Court of Appeals for the District of Columbia Circuit. Simultaneous with the promulgation of these eight MCLGs, the EPA proposed MCLGs for a multitude of other contaminants, including VOCs, inorganic chemicals, and microbiological contaminants.

In 1986, Congress acted to speed up the SDWA standard-setting process. First, the EPA was granted the authority to promulgate MCLGs and MCLs at the same time. Second, the EPA was placed on a mandatory schedule. For the MCLGs proposed and promulgated and the MCLs proposed in 1985, the EPA was given up to 12 months to promulgate MCLGs and MCLs for 9 contaminants; up to 24 months for 40 more contaminants; and up to 36 months for the additional 33 contaminants. The EPA would be able to make up to 7 substitutions—to be determined within 12 months of enactment—for contaminants on the list. As for future standards, by January 1, 1988, and at 3-year intervals thereafter, the EPA must have published a list of contaminants that are known or anticipated to occur in public water systems and that may have any adverse effect on the health of persons. By January 1, 1991, the EPA must issue MCLGs and MCLs for at least 25 of these contaminants. Moreover, within 18 months of enactment, the EPA must have promulgated regulations requiring public water systems to monitor for listed unregulated contaminants at least once every 5 years.

The 1986 SDWA amendments addressed the question of whether public water systems with VOCs problems should be required to install granular activated carbon (GAC) as a treatment technique. The EPA had proposed this in 1978, but the proposal was withdrawn after vehement protests from water purveyors. The amendments recognize that GAC "is feasible for the control of synthetic organic chemicals, and any technology, treatment techniques, or other means found to be the best available for the control of synthetic organic chemicals must be at least as effective in controlling [them] as granular activated carbon."

Two new regulatory programs were authorized by the 1986 amendments. The first involves filtration and disinfection. Within 18 months of enactment, the EPA must have promulgated regulations specifying criteria under which filtration (including coagulation and sedimentation, where appropriate) is required as a treatment technique for public water systems supplied by surface water sources. Raw water quality, protection offered by watershed management, and treatment practices were to be considered in developing the criteria. Within 30 months after these regulations were promulgated, states with primary enforcement responsibility must decide which surface water utilities need to filter water to protect against bacterial and viral contamination. The EPA will make these determinations in other states. Utilities would have another 18 months to install needed filtration facilities. As for disinfection, the EPA has 36 months after enactment to promulgate regulations requiring utilities to disinfect their water.

The second new regulatory program deals with lead in water pipes. Beyond insignificant allowable amounts, the use of lead in installing or repairing water pipes is prohibited. Utilities with lead in their distribution systems must notify customers of the sources of lead in the system, potential adverse health effects, reasonably available mitigation measures, and any necessity for seeking alternative water supplies. This program is to be administered by the states.

The EPA and states may grant variances and exemptions from primary regulations. Variances for unspecified lengths of time may be issued to public water systems that, despite the application of best available technology economically achievable, cannot comply with primary regulations because of impurities in their raw water sources. Exemptions of up to three years may be granted if ''due to compelling factors (which may include economic factors), the public water system is unable to comply.'' Neither a variance nor an exemption may be granted where there would be ''an unreasonable risk to health.''

Similar to its behavior with regard to POTWs under the Clean Water Act, the EPA's enforcement of the SDWA was poor for a decade. There were a number of reasons for this failure. Enforcement against public water systems is subject to the same infirmities as enforcement against POTWs because legislatures, courts, and agencies are reluctant to impose heavy fines and jail sentences on public bodies and their officials. As a result, the 1974 SDWA tied the EPA's hands by (1) not authorizing the EPA to impose administrative compliance orders or penalties; (2) limiting court-imposed civil penalties to $5,000 per day; and (3) limiting imposition of civil penalties to ''willful'' violators. In 1986 Congress gave the EPA more powerful enforcement weapons. The EPA was given the authority to issue administrative compliance orders and impose administrative penalties of $5,000 or less. The EPA must issue a compliance order or commence a civil action in court if a state has not taken enforcement action within 30 days after an EPA notice of noncompliance. Maximum civil penalties were raised from $5,000 to $25,000 per violation, and violations need no longer be ''willful'' in order to be actionable. Tampering with, attempting to tamper with, or threatening to tamper with a public water supply were made crimes.

The SDWA contains a unique enforcement mechanism—mandatory public exposure of violators. All violators of the SDWA or a comparable state statute must publish notices of violation in local newspapers and include violation notices with water bills. Citizens can use this information to work for compliance through local political channels or to bring citizen suits for injunctions. Once again, this system presupposes a high degree of citizen vigilance and litigation capability. Moreover, the public notice requirements were more honored in the breach than the observance. In the 1986 amendments, Congress modified the notice requirements to make them more realistic and more sensitive to the frequency and severity of violations. It also authorized a civil fine of up to $25,000 for violations of the notice requirements.

A number of states have gone further than the SDWA in protecting drinking water. ''Proposition 65'' in California requires the preparation of a list of carcinogens and reproductive toxins and then a prohibition on discharges to drinking water of those substances. In New Jersey, ''A-280'' establishes a Drinking Water Quality Institute to recommend MCLs for 22 listed chemicals and any other dangerous chemicals found in the state's drinking water. The state agency must then promulgate MCLs which ''with respect to carcinogens, permit cancer in no more than one in a million persons ingesting that chemical for a lifetime," and for all other chemicals and carcinogens with public health benefits, eliminate all adverse effects "within the limits of practicability and feasibility." If one of these MCLs is exceeded, the public water system has a maximum of one year to achieve compliance.

SALINE WATER CONVERSION

Since 1952 the Department of the Interior has been conducting a research program for the development of processes for economically converting saline water into fresh water. The present Saline Water Conversion Program was authorized by Title II of the Water Research and Development Act of 1978. The GAO has concluded that "a practical, low-cost desalting method has not been achieved;"[35] and the GAO recommends that the program be thoroughly reevaluated.

Chapter 40
THE COMMON LAW
AND WATER QUALITY

This Part has justifiably concentrated on the statutory law of water treatment. However, the common law still plays an important role in achieving and maintaining clean water in the United States. Common law litigation is generally the first response of our legal system to an emerging issue. Then, as the issue evolves and grows more complex, municipal ordinances and state legislation are brought to bear on the problem. Finally, federal legislation establishes the partnership between federal and state administrative agencies that is characteristic of a mature regulatory system responding to a significant national issue.

The replacement of common law actions for injunctions and damages by administrative regulation is inevitable because of severe limitations on the common law as a pollution control strategy. Private litigation is sporadic in place and time, while administrative regulations are universally applicable and often uniform in effect. Second, many private actions are inhibited by restrictive rules of standing and the expense of bringing litigation. On the other hand, administrative agencies automatically possess standing and agencies, with their professional enforcement staffs, are far less sensitive to expense than the average citizen. Third, in deciding whether an injunction should be granted, courts "balance the equities" between the interests of a plaintiff and society as a whole. The balance is more likely to be in favor of pollution control where a public agency, also representing society as a whole, is the plaintiff. Finally, a plaintiff in an environmental common law action is inherently at a grave disadvantage because he must prove, by a preponderance of the evidence, that a particular discharger, or group of dischargers acting independently, caused his injury. The difficulty of fulfilling this burden under the conditions of scientific uncertainty that characterize environmental problems is illustrated by the 15-year struggle to curtail Reserve Mining Company's discharge of taconite tailings into Lake Superior.[36]

Although they cannot bear the brunt of modern pollution control, common law actions still play an important part in regulatory systems. Lawsuits based on legal theories such as nuisance, trespass, negligence, abnormally dangerous activities, and interference with riparian rights "fill in the gaps" left by water pollution control statutes and regulations. One such gap is the failure of pollution control statutes to provide for compensating victims of illegal discharges (i.e., "third-party recoveries"). Common law damage recoveries not only achieve compensation but also serve as an additional deterrent to illegal discharges. Another gap is the inability of federal and state officials to take enforcement action against dischargers who are in compliance with their discharge permits but are still causing injury to others. For these reasons, most of the federal statutes discussed in this part include "savings clauses" similar to section 505(e) of the Clean Water Act: "Nothing in this section shall restrict any right which any person (or class of persons) may have under any statute or common law to seek enforcement of any effluent standard or limitation or to seek any other relief."

The case of *Stoddard* v. *Western Carolina Regional Sewer Authority*[37] is an example of a common law damage recovery. There, three South Carolina lakeshore property owners sued a sewer authority that was discharging into the lake on which they owned property. Plaintiffs asserted two causes of action: (1) that defendant was violating its Clean Water Act discharge permit and should be fined; and (2) that defendant was liable for a continuing nuisance that had ripened into a "taking" of their property, and should therefore be made to pay them compensation.

The treatment plant was built in 1967, and in 1977 it was granted a discharge permit by the appropriate state agency. The plant was designed to treat only domestic waste, had minimal equipment, no chlorination process, and no means of sludge disposal. Over the years, the proportion of high-strength industrial waste treated at the plant increased until it contributed 85% of the plant's total organic load. Although the sewer authority had the power to order its contributing industries to pretreat their wastes, it did not do so until 1984.

From 1979 until 1983, the plant was operated "as if the NPDES permit did not exist." During this time, the plant regularly exceeded its permit limits for BOD, TSS, DO, pH, and fecal coliform. In fact, the plant "went septic" twice. The frighteningly high fecal coliform levels in the discharge could have been prevented by the installation of a $50,000 chlorinator, which defendant did not procure until the state agency ordered it to do so in 1983.

Between 1979 and 1983, the agency had issued numerous notices of violation to the defendant, characterizing the plant's operation as "unacceptable."

Plaintiff's lake was heavily impacted by the defendant's illegal discharges. Fish kills occurred twice. Premature euthrophication was rampant. Frequently, odors were so noxious that people could not remain in the vicinity; algae clumps and human sewage often floated on the surface of the lake. Even after pollution of the lake had stopped—and dissolved oxygen violations were still occurring at the time of trial—the lake would not recover for at least five years.

The federal district court rejected defendant's claims of sovereign immunity, holding that, under South Carolina Law, a public body cannot be granted immunity from a nuisance that amounts to a constitutional taking. Finding that a taking had occurred, the Court awarded one plaintiff $198,000, the second $36,000, and the third $63,900. However, the district court refused to fine defendant based on plaintiffs' citizen suit cause of action.

The Fourth Circuit Court of Appeals upheld the damage award, and also ordered the district court judge to levy some fine against the defendant. This victory on their citizen suit claim enabled plaintiffs to recover their legal fees from the defendant.

The case of *Village of Wilsonville* v. *SCA Services, Inc.*[38] exemplifies the situation where defendant is in compliance with a permit but is nevertheless causing environmental damage. In *Wilsonville*, the Illinois Supreme Court affirmed a trial court judgment ordering the defendant to close its hazardous chemical waste landfill in Wilsonville, to remove all toxic waste and contaminated soil at the disposal site, and to restore and reclaim the site. The court was impressed by evidence that groundwater contamination was inevitable, despite engineering techniques, because the site was located over an abandoned coal mine that was already settling and causing subsidence at the landfill site. Moreover, there was convincing testimony that hazardous materials had been buried in such a sloppy manner that an explosion was likely if oxygen could reach the chemicals. The court rejected the defendant's argument that the landfill site and each batch disposal had been permitted by the Illinois Environmental Protection Agency, because IEPA had granted the permits based on the defendant's inaccurate data. The U.S. Environmental Protection Agency appeared as *amicus curiae* (friend of the court) and filed affidavits claiming that removal and transportation of wastes from the site would be riskier than allowing the site to remain open and operate in a "clean" mode, but the court could not understand how safe operation was possible in light of the subsidence. The court acknowledged the need for hazardous waste disposal sites, but declared that they must be located in suitable locations where they will not pose a threat to public health. In "balancing the equities" the Illinois Supreme Court firmly held that in Illinois the rights of a private landowner are presumptively superior to the public benefit or convenience from having a business operate at a particular location.

Private litigation such as *Stoddard* and *Wilsonville* allows our legal system to go beyond statutes and regulations in protecting the public against pollution. In addition, common law actions call attention to issues that legislatures may not have adequately addressed. For example, *Wilsonville* emphasizes the need for hazardous waste facility-siting legislation in order to forestall building hazardous waste sites over abandoned coal mines. The decision deals with water resources and land use in a comprehensive manner. This enlightened approach should be taken by legislatures, administrative agencies, and other courts in addressing future water resource problems.

Chapter 41
REGIONAL LAND-USE CONTROLS

At a certain stage of land development, water pollution will inevitably occur, regardless of technological or managerial expedients used to control it. The greater the concentration of industry in a particular area, the greater will be the likelihood of pollution caused by spills, breakdowns, and illegal discharges. The greater the concentration of residential development in a particular area, the greater will be the likelihood of pollution from stormwater runoff over impervious surfaces. Consequently, controls on land development must be imposed if a critical water-related resource is to be protected.

In the United States, most controls on land development are imposed at the local level. However, many of our precious and vulnerable water resources transcend municipal boundaries. Thus, regional land-use controls are becoming a prominent feature of our regulatory landscape. Three examples of this trend are New York's Adirondack Park Agency, Florida's Environmental Land and Water Management Act, and California and Nevada's Tahoe Regional Planning Agency. Then, the concept of watershed management will be briefly explored.

ADIRONDACK PARK AGENCY

The Adirondack Park encompasses approximately six million acres in northern New York State.[39] Established in 1892, nearly all its state-owned lands have been protected since 1895 by a clause of the New York State Constitution that declares them "forever wild." State-owned and nonstate-owned lands are intermingled to a significant degree in the park. About 38% of the land is owned by the State, while the remainder is divided into parcels of diverse sizes owned by individuals, corporations, institutions such as universities and camps, and municipalities.

Among its outstanding natural resources, the park contains the headwaters of five major water basins. More than 1,200 miles of Adirondack rivers have been designated under New York State's Wild, Scenic, and Recreational Rivers System Act.

In the face of potential development pressure on the Adirondacks, the New York State Legislature, in 1971, passed the Adirondack Park Agency Act, creating the Adirondack Park Agency and requiring it to prepare a comprehensive management plan for the park's state lands and a land-use and development plan for the nonstate lands in the Adirondacks.[39] The so-called "Private Land Plan" was enacted as a statute in 1973.

The Private Land Plan divides nonstate lands into six land-use areas: (1) hamlets; (2) moderate-intensity-use areas; (3) low-intensity-use areas; (4) rural-use areas; (5) resource management areas; and (6) industrial-use areas. The Plan sets compatible uses and overall density limitations for each land-use area. For example, only 15 residential units per square mile are allowed in a resource management area. Private lands are frequently zoned in a low-density area because of their proximity to state lands.

These restrictions are implemented through a regional project review system. Except for the act's shoreline development restrictions, which apply to all development in the park, the Agency has jurisdiction only over new projects of regional significance. Each land-use area has its own corresponding set of regional projects and critical environmental areas. Regional projects are divided into Class A and Class B regional projects. Class A projects—large projects with serious environmental impacts—must be approved by the Agency. Class B projects may be approved by a municipality whose land-use controls have been accepted by the Agency. Each land-use area has its corresponding Class A and Class B regional projects.

According to Professor Booth,[39] the Adirondack Park Agency has been generally successful in protecting the park's natural resources, even though it has been confronted by strong local opposition. Ironically, the agency's major failure is that its universally applicable shoreline restrictions are inadequate to protect the park's shorelines.

ENVIRONMENTAL LAND AND WATER MANAGEMENT ACT

Florida's high rate of growth has led to serious water supply and water quality problems. After a major drought in 1971, the legislature enacted the Environmental Land and Water Management Act of 1972. This act creates a partnership among state, regional, and local institutions to deal with development in areas of critical state concern and developments of regional impact.[40]

An "area of critical state concern" is (1) an area containing or having a significant impact upon environmental, historical, natural, or archeological resources of regional or statewide importance, (2) an area significantly affected by or having a significant affect upon an existing or proposed major public facility or other area of major public investment, or (3) a proposed area of major development potential. The Division of State Planning recommends areas for designation as areas of critical state concern to the governor and his six cabinet members, all of whom are elected. If this administrative commission accepts the recommendation, the state legislature must then approve the designation. Only 5% of the state's total land area may be designated as critical. But the legislature may exempt an area from the 5% rule, as it has done with the Big Cypress Swamp.

Once an area of critical state concern has been designated, local governments must prepare land development regulations for it in accordance with state guidelines. If a local government does not adopt adequate regulations, the Division of State Planning must prepare regulations and submit them to the administrative commission for adoption. Local governments must enforce land-use controls in critical areas. If a local government does not adequately enforce the applicable controls, the state may sue to block offending development.

A "Development of Regional Impact" (DRI) is defined as "any development which, because of its character, magnitude, or location, would have a substantial effect upon the health, safety, or welfare of citizens of more than one county." DRIs include airports, attractions and recreation facilities, electrical generating facilities and transmission lines, hospitals, industrial plants and industrial parks, mining operations, office parks, petroleum storage facilities, port facilities, residential developments, schools, and shopping centers.[40]

A developer of a DRI must file an application for development approval with the local government having jurisdiction. The local government then submits the application to the appropriate regional planning agency. The regional planning agency is required to review the application and submit a report to the local government. In preparing its report and recommendations, the regional planning agency must consider and balance the favorable and unfavorable impacts of the proposed development on (1) the environment and natural resources, (2) the economy, (3) public facilities, (4) public transportation, and (5) housing. The local government must consider this report in making its decision to approve the proposed DRI, to deny it, or to conditionally approve it. The local government's decision may be appealed to the governor and his cabinet by the regional planning agency, the Division of State Planning, or the developer. Finally, judicial review is available.

LAKE TAHOE REGIONAL PLANNING AGENCY

The problems plaguing Lake Tahoe have been vividly described by Professor Williams:

> Lake Tahoe, one of the few large high-altitude lakes in the world, is surrounded by steep slopes with shallow soils, and has one shore in California and the other in Nevada. . . . In this situation, the inevitable result is that any substantial amount of development is accompanied by heavy soil erosion, with little opportunity for revegetation. Since the 1960's Lake Tahoe has been developing rapidly as a mountain resort, devoted primarily to gambling. The environmental consequences have been predictable enough; erosion has been widespread, and the famous blue of the lake has begun to fade. . . . The basic problem at Lake Tahoe is the almost total lack of support for environmental-protection measures; in fact, there has been widespread and strong local resistance.[41]

Local resistance to environmental protection has been exacerbated by the interstate nature of the problem.

In 1970, Congress approved an interstate compact establishing a bistate Tahoe Regional Planning Agency. This Agency, according to the original compact, could prevent a proposed development only by a negative majority vote of each delegation—giving both states a veto over development disapprovals. This virtually paralyzed the Agency, because the Nevada delegation "was normally solidly pro-development (any development), particularly where gambling was involved . . . ."[41] Thus, a land-use plan prepared by the Agency's staff was vitiated, and variances for casino-hotels were common. California became so frustrated by this impasse that it formed its own Regional Planning Agency affecting only the area in California. This agency has exercised strict land-use controls within its limited jurisdiction. Recent amendments to the Tahoe compact were intended to strengthen the interstate agency's regulatory powers, but, according to Professor Williams, it is still unlikely that any development will be disapproved.

Three models of regional land-use controls to protect water resources have been presented. The "regional commission" model is represented by the Adirondack Park Agency. Two other successful regional commissions are New Jersey's Pinelands Commission and California's San Francisco Bay Conservation and Development Commission. The Adirondack Park Agency has been able to overcome local opposition because of the nearly 100-year tradition of conservation in the Adirondack Park and because of strong support from the state legislature. The "state-regional-local partnership" model, as embodied in the Florida Environmental Land and Water Management Act, has succeeded because of the almost universal appreciation in Florida that the state's water resources are under extreme and increasing stress. In contrast, the recognition of a water resources crisis has not occurred on the Nevada side of Lake Tahoe. The "cooperative regulatory" model has failed there because the two states involved have widely different goals and attitudes toward growth management.

WATERSHED MANAGEMENT

Another variety of regional land-use management in the interests of water resources management is "watershed management." This concept has been analyzed and recommended since the New Deal, but except for the Tennessee Valley Authority and some watersheds in public ownership, watershed management has not taken hold. The reason for this is political. American political jurisdictions have not been established with regard to watershed boundaries. And the obstacles to interjurisdictional land-use control are most often insurmountable. But, as the North Carolina example in Chapter 34 illustrates, watershed management is possible where there is a widespread perception of crisis, where a respected planning agency exists, and where the management plan is equitable.

REFERENCES FOR PART IV

1. Dworsky, L. B. *Water and Air Pollution* (New York: Chelsea House Publishers and Van Nostrand Reinhold Co., 1971), p. 33.
2. 33 U.S.C. 1251 et seq.
3. Rose-Ackerman, S. "Market Models for Water Pollution Control: Their Strengths and Weaknesses," *Public Policy* 25(3):383–406 (1977).
4. Benstock, M., and L. Zwick. *Water Wasteland* (New York: Grossman Publishers, 1971), ch. 14.
5. "Many Water Quality Standard Violations May Not Be Significant Enough to Justify Costly Preventive Actions," USGAO Report CED-80-86 (1980), p. 6.
6. Ackerman, B. A., S. Rose-Ackerman, J. W. Sawyer Jr., and D. W. Henderson, *The Uncertain Search for Environmental Quality* (New York: The Free Press, 1974), Part IV.
7. *EPA* v. *California* 426 U.S. 200, 48 L. Ed. 2d 578, 96 Sup. Ct. 2022 (1976).
8. "A Legislative History of the Federal Water Pollution Control Act Amendments of 1972 and the Clean Water Act of 1977," Congressional Research Service, U.S. Government Printing Office (1973, 1978).
9. 40 CFR § 131.1 et seq.
10. *DuPont* v. *Train,* 430 U.S. 112, 97 S. Ct. 965, 51 L. Ed. 2d 204 (1977).
11. Gaba, J. M. "Federal Supervision of State Water Quality Standards Under the Clean Water Act," *Vanderbilt Law Review* 36: 1167–1219 (1983), pp. 1213–1219.
12. Downey, D., and S. Sessions, "Innovative water quality-based permitting: a policy perspective," *Jour. of the Water Poll. Con. Fed.* 57:358–365 (1985), p. 360.
13. "How stringent should marine waivers from secondary treatment be?" *Jour. of the Water Poll. Con. Fed.* 58:1101–1104 (1986), p. 1103.
14. "EPA's Innovative Technology Program For Waste Water Treatment Needs Better Controls," USGAO Report CED-84-79 (1984).
15. "Costly Wastewater Treatment Plants Fail to Perform as Expected," USGAO Report CED-81-9 (1980), p. 15.
16. "User Charge Revenues for Wastewater Treatment Plants—Insufficient to Cover Operation and Maintenance," USGAO Report CED-82-1 (1981).
17. "A New Approach Is Needed for the Federal Industrial Wastewater Pretreatment Program" USGAO Report CED-82-37 (1982), p. 14.
18. *NRDC* v. *EPA,* 790 F. 2d 289 (1986).
19. Wall, T. M. and R. W. Hanmer, "Biological testing to control toxic water pollutants," *Jour. of the Water Poll. Con. Fed.* 59:7–12 (1987), p. 8.
20. *Weinberger* v. *Romero-Barcelo,* 456 U.S. 305, 102 S. Ct. 1798, 72 L. Ed. 2d 91.
21. "More Efficient Action by the Environmental Protection Agency Needed to Enforce Industrial Compliance with Water Pollution Control Discharge Permits," USGAO Report CED-78-182 (1978).

22. "Wastewater Dischargers Are Not Complying with EPA Pollution Control Permits," USGAO Report RCED-84-53 (1983).

23. Blaser, Zeni, and Company. "Role of Interstate Water Pollution Control Commissions," NTIS PB-257 759 (1975), p. 16.

24. 25 ERC 1457 (1987).

25. Goldfarb, W. "Water Quality Management Planning: The Fate of 208," *University of Toledo Law Review* 8:105–134 (1976).

26. Blaha, D. W. "Innovations In Watershed Protection: A North Carolina Case Study," *Environmental Planning Quarterly* 4:10–12 (1986).

27. "Problems and Progress in Regulating Ocean Dumping of Sewage Sludge and Industrial Wastes," USGAO Report CED-77-18 (1977), pp. 17–25.

28. Tripp, J. T. B., and A. B. Jaffe. "Preventing Groundwater Pollution: Towards a Coordinated Strategy to Protect Critical Recharge Zones," *Harvard Environmental Law Review* 3:1–47 (1979), pp. 1,2.

29. Dycus, S. J. "Development Of A National Groundwater Protection Policy," *Boston College Environmental Affairs Law Review* 11:211–271 (1984).

30. *Protecting the Nation's Groundwater From Contamination* (Washington, D.C.: U.S. Congress, Office of Technology Assessment, OTA-0-233, October 1984), p. 45.

31. Gordon, W. A. *Citizen's Handbook On Groundwater Protection* (New York City: Natural Resources Defense Council, Inc., 1984), p. 109.

32. *Survey Of State Ground-Water Quality Protection Legislation, 1985* (Washington, D.C.: USEPA, 1987), pp. 5, 6.

33. *Groundwater Protection* (Washington, D.C.: The Conservation Foundation, 1987), p. 173.

34. Trauberman, J. "Oil and Hazardous Spill Control," in *Air and Water Pollution Control Law* (Washington, D.C.: Environmental Law Institute, 1982), p. 592.

35. "Desalting Water Probably Will Not Solve the Nation's Water Problems, But Can Help," USGAO Report CED-79-60 (1979), p. i.

36. Bartlett, R. V. *The Reserve Mining Controversy* (Bloomington: Indiana University Press, 1980).

37. 784 F. 2d 1200 (1986).

38. 426 N.E. 2d 824.

39. Booth, R. "New York's Adirondack Park Agency" in Brower, D. J., and D. S. Carol, Eds. *Managing Land-Use Conflicts* (Durham, NC: Duke University Press, 1987), p. 140.

40. Natural Resources Defense Council. *Land Use Controls in the United States* (New York: The Dial Press/James Wade, 1977), p. 283.

41. Williams, N. *American Planning Law* (5 vols; Wilmette IL: Callaghan & Company, 1985), 5, p. 796.

PART V

Mediation of
Water Resources Disputes

NEED FOR MEDIATION

Disputes over the allocation of water resources have traditionally been settled by administrative agencies, courts, and legislatures, sometimes singly
but more often in combination. For example, proponents and opponents
of a controversial federally-funded impoundment will—either successively
or simultaneously—make their cases to (1) federal agencies such as the Corps
of Engineers, Bureau of Reclamation, Federal Energy Regulatory Commission, and Environmental Protection Agency; (2) federal courts to review
agency decisions; and (3) congressional committees involved with authorizing and appropriating funds for the project.

However, significant dissatisfaction has arisen with regard to all phases
of this process. Administrative agencies are frequently viewed as inaccessi-
ble to those who lack the money and expertise to conduct effective administrative lobbying, "captured" by their clientele in the regulated community,
and politically illegitimate because of their insulation from the electorate.
Litigation is perceived as expensive, slow, unnecessarily adversarial, and
preoccupied with technicalities at the expense of the "real issues." Legislation is also seen as expensive and slow; moreover, legislative compromise
often results in vague, "symbolic" declarations that postpone disputes to
the implementation stage rather than resolving them.[1]

In this climate of disenchantment with traditional legal dispute-
resolution mechanisms, there has been a good deal of experimentation with

voluntary environmental dispute-resolution processes such as mediation. "Since 1973, when two mediators were invited by the governor of Washington State to help settle a dispute over a proposed flood control dam [on the Snoqualmie River], mediators have been employed in over 160 environmental disputes in the United States."[2] Many of these mediated environmental disputes have involved water resources.

MEDIATION: ADVANTAGES AND DISADVANTAGES

Mediation is one type of "environmental dispute resolution" (EDR), also referred to as "alternative dispute resolution" (ADR). In contrast to administrative decisionmaking, litigation, and legislation,

> (m)ediation is a voluntary process in which those involved in a dispute jointly explore and reconcile their differences. The mediator has no authority to impose a settlement. His or her strength lies in the ability to assist the parties in resolving their own differences. The mediated dispute is settled when the parties themselves reach what they consider to be a workable solution.[3]

The voluntariness of the mediation process is thought to guarantee the legitimacy and the implementability of mediated resolutions. And despite a lack of empirical evidence, mediation is widely believed to be cheaper and faster than formal public modes of dispute resolution.

But the extensive literature on environmental mediation is replete with warnings that mediation is not a panacea that will replace the administrative process, litigation, or legislation.

> Voluntary dispute resolution processes may not always be appropriate, however, and they may function better as supplements to legislative, administrative, or judicial processes. Disputes over environmental issues are so varied that no one dispute resolution process is likely to be successful in all situations. Depending on the circumstances and what each party hopes to achieve, the parties may prefer to litigate, to lobby for legislative change, to turn to an administrative agency, to negotiate a voluntary resolution of the issues with one another, or engage in some combination of these options.[2]

In fact, experienced environmental mediators have concluded that mediation is only appropriate in about ten percent of environmental disputes.[1,3]

Once disputants agree to enter into mediation they almost always reach an agreement, and the provisions of the agreement will probably be implemented.[2] But convincing the disputants to undertake mediation is extraordinarily difficult.

A central tenet of mediation theory is that "mediation cannot work as long as one or more parties to the dispute holds out a hope of 'winning' eventually, without compromise, or when at least one party perceives a benefit to be derived from indefinite continuation of the dispute."[3] In other words, disputants will only negotiate in good faith if they believe that they cannot win through other strategies. In order for mediation to be a viable option, both a stalemate and a relative balance of power must exist.[1] The disputants must feel frustrated by their inability to win everything, and fearful of losing everything by the uncontrollable action of a court, legislature, or agency. Mediation becomes attractive in this "high risk situation."[1] Thus, inconclusive litigation is frequently a precursor of mediation. The "last resort" quality of mediation, more than any other factor, accounts for its reputation as a "ten percent solution."

Moreover, value disagreements—conflicts over moral principles— cannot be mediated. Vehement critics of a proposal to dam a scenic river will not enter mediation over mitigating the dam's environmental impacts. Mediation functions best when "there is a basic consensus over environmental values and goals and the dispute centers around different approaches to achieving them."[1] Mediation generates compromise solutions, and one of its goals "is to encourage responsible, planned, and progressive development that will preserve and enhance the natural and cultural environment of the earth."[4] There is no tolerance for "all or nothing" positions in the mediation process. Matters of principle are best left to courts and legislatures.

A related shortcoming of mediation is its relative inapplicability to broad, generic policy issues, as opposed to site or regulation-specific controversies.[3] There have been a few successful policy mediations,[2] but mediated agreements involving particular projects and sets of administrative regulations are far more common. Broad policy questions often reflect nonmediable value differences. Moreover, identifying, organizing, and mediating among all the groups interested in, for example, federal groundwater policy would be an overwhelming task. Refining the issues and determining necessary data would also be unmanageable. Indeed, the "smaller" the conflict—in terms of geographical area affected, number of issues involved, and locus of political decisionmaking—the more mediable it will be.[4]

THE MEDIATOR'S ROLE

Where other alternatives are unattractive, where the dispute is not primarily over values, and where the controversy is one of manageable proportions, mediation can be a valuable option. It has been particularly successful where the dispute is based on personality conflict, conflicting data, and conflicting positions that conceal common or consistent interests:

> Mediators commonly ask the various parties, either in confidence or at the bargaining table, the reasons why they hold a particular position—that is, what

their underlying objectives or interests in the dispute are. In doing this, the mediator hopes to discover that those interests are not as incompatible as the parties' positions on the issues appear to be—either because they can find some common ground or because their interests are sufficiently *different* that they are not mutually exclusive—thereby giving the parties a new opportunity to develop alternatives that may satisfy their real needs more effectively.[2]

When interests are found to be inherently inconsistent, negotiated compensation is still a possibility.[1]

Once disputants have committed themselves to mediation, the success or failure of the effort will, above all, depend on the skill and objectivity of the mediator. An effective mediator will ensure that: (1) all the parties that have a stake in the outcome of the dispute are identified; (2) the relevant interest groups are adequately represented; (3) a climate of trust and willingness to negotiate has been established; (4) the parties agree on the scope of issues to be negotiated; (5) the best available information and expertise is brought to the negotiation; (6) fundamentally different values and assumptions are confronted; (7) a sufficient number of possible solutions or options are developed; (8) a reasonable deadline for completion of negotiations is set; (9) the weighting, scaling, and amalgamations of judgments about costs and benefits are undertaken jointly; (10) fair compensation and mitigating actions are negotiated; (11) any bargains struck are legal and economically feasible; (12) relevant public agencies with decisionmaking authority either participate in the negotiations or agree to implement accords; and (13) all parties implement any agreements reached among them.[2,3,5]

The skillful mediator will foster an atmosphere of cooperative, creative problem-solving rather than one of hostile confrontation. The disputants will be encouraged to take control of the situation by voluntarily resolving their differences instead of allowing a public official to decide the issue in a way that might produce an unpredictable "Win-Lose" result. The mediator's goal is to develop an option that may not entirely satisfy any single disputant, but nevertheless leaves all disputants better off than they were at the outset of negotiations—that is to say, a "Win-Win" result.

MEDIATION AS COOPTATION

Critics of environmental mediation are concerned that "(i)ts very informality and lack of structure make it open to exploitation by unethical participants."[1] Seduction and cooptation may occur because environmental mediation lacks the procedural safeguards of a judicial proceedings and the ongoing relationship characteristic of a labor mediation. Environmentalists may be deficient in negotiation experience, access to information, resources to pursue other options if the mediation fails, and the time to negotiate extensively.[1] Their justifiable anger may be dispelled by the symbolic assurance of an ostensibly rational process that is, in reality, being exploited to achieve emotional pacification. And mediators may be more interested in resolving disputes than in fair settlements.

These dangers are exacerbated where a power disequilibrium exists among the parties to a mediation. Contrary to popular belief, most of the mediated environmental disputes have involved governmental agencies.[2]

> Perhaps the most troublesome aspect of government-mandated or sponsored mediation efforts is the possibility that they are being used simply as a way for the government to more efficiently implement its policies. At first glance, government-sponsored mediation efforts appear to constitute a desirable new form of citizen and interest group participation in environmental policymaking. The government looks willing to sit down and negotiate over important issues. But . . . public participation is hardly the motive behind these efforts. What the government is primarily interested in . . . is reducing court costs and speeding compliance with its policies.[1]

In addition to its superior resources, government has the ultimate power in a mediation—the power to implement or not implement an agreement.

These warnings that the mediation process may be manipulated to coopt weaker opponents must be taken seriously. But the essential voluntariness of mediation provides a safeguard against seduction. In order to ensure a stable agreement, negotiators must convince members of their own organizations and ideologically-related groups unrepresented in the negotiations that the settlement is in their best interests. Otherwise, dissidents may turn to litigation or legislative lobbying in an attempt to subvert the agreement. Unequal access to litigative resources is mitigated somewhat by the existence of public-interest law firms and statutory provisions that permit successful parties to recover reasonable attorney and expert witness fees. Legislative lobbying by environmentalists is considerably facilitated by the prevailing public consensus for environmental protection, media interest in the environment, and the resulting sensitivity of legislators to environmental issues.

MEDIATION OF WATER RESOURCES DISPUTES

Much of the literature on environmental mediation consists of case studies; but none of these will be explored here because (1) most published case studies are "heavy on description and light on analysis";[1] (2) the principles that can be extracted from the studies have been enunciated above; (3) each dispute is unique; and (4) mediation is an art rather than a science, and a successful mediator's blend of trustworthiness, objectivity, and facilitation skill is, to a great extent, indescribable.

Nevertheless, the reader may find the following list of mediated water resource dispute case studies useful in order to appreciate the variety of water resources disputes that have been mediated and, perhaps, to find an experienced mediator. Each specific dispute is identified by subject matter and is followed by the sources of published case studies and the location of the controversy:

- —Basin Management (nutrient control)[2] (Patuxent River, Maryland).
- —Basin Management (flood control-recreation-dam construction)[2,4,5] (Snoqualmie River, Washington).
- —Basin Planning (port development)[2] (Columbia River Estuary, Oregon).
- —Basin Planning (small watershed project)[2] (western Massachusetts).
- —Compensation for Water Pollution[6] (Ontario, Canada).
- —Dam Repair[2] (Jackson County, Wisconsin).
- —Dredging[2] (White Oak River, North Carolina).
- —Indian Fishing Rights[2] (Lake Superior, Wisconsin).
- —Indian Water Rights[2,7] (Papago Tribe, Arizona).
- —Port Facilities[8] (Port Townsend, Washington).
- —POTW Siting (growth management)[9] (Jackson, Wyoming).
- —Pumped Storage Facility[2,8] (Hudson River, New York).
- —River Recreation[2] (Rogue River, Oregon).
- —Sludge Composting Facility[2] (Barrington, Rhode Island).
- —Small Hydropower Project[2] (Middlebury, Vermont).
- —Small Hydropower Project[2] (Swanville, Maine).
- —Water Supply Hookups[2] (Fitchburg, Wisconsin).
- —Water Supply Hookups[2] (Charles City County, Virginia).
- —Water Supply (dam construction)[2,5,9,10] (Denver, Colorado).
- —Water Pollution Control (heat from nuclear plant)[2] (Lake Anna, Virginia).
- —Water Pollution Control (cheese wastes)[2] (Franklin County, Vermont)
- —Water Pollution Control (taconite from iron ore refining)[4] (Lake Superior, Wisconsin).
- —Water Pollution Control (PCBs)[4] (Hudson River, New York).

Mediation could undoubtedly be utilized to resolve more water resource disputes than has been the case in the past. It should seriously be considered where an impasse has been reached in a controversy that is not fundamentally over values. At some point in the future, when additional data become available about conflict resolution, it may be possible to predict where and when a judicial or legislative impasse will occur, thereby obviating protracted and costly litigative and legislative efforts that are doomed to failure. But it is a mistake to make mediation compulsory, as some states have done with regard to siting of solid and hazardous waste facilities. In voluntariness lies the legitimacy and the stability of mediated agreements, especially where government is one of the disputants. Compulsory mediation—as opposed to compulsory arbitration—is a contradiction in terms.

REFERENCES FOR PART V

1. Amy, D. J. *The Politics of Environmental Mediation* (New York: Columbia University Press, 1987), p. 27.
2. Bingham, G. *Resolving Environmental Disputes* (Washington, D.C.: The Conservation Foundation, 1986), p. 1.
3. RESOLVE, Center for Environmental Conflict Resolution, *Environmental Mediation: An Effective Alternative?* (Palo Alto, CA, 1978), p. 2.
4. Mernitz, S. *Mediation of Environmental Disputes* (New York: Praeger Publishers, 1980), p. 163.
5. Susskind, L. "Environmental Mediation and the Accountability Problem," *Vermont Law Review* 6(1):1–47 (1981), p. 14.
6. West, L. "Mediated Settlement Of Environmental Disputes: Grassy Narrows and White Dog Revisited," *Environmental Law* 18(1):131–150 (1987).
7. Folk-Williams, J. A. "The Use of Negotiated Agreements to Resolve Water Disputes Involving Indian Rights," *Natural Resources Journal* 28(1):63–103 (1988).
8. Talbot, A. R. *Settling Things* (Washington, D.C.: The Conservation Foundation, 1983), pp. 79–89.
9. Susskind, L., Bacow, L., and Wheeler, M., eds., *Resolving Environmental Regulatory Disputes* (Cambridge, MA: Schenkman Books, Inc., 1983), pp. 86–121.
10. Carpenter, S. L. and Kennedy, W. J. D. "The Denver Metropolitan Water Roundtable: A Case Study in Reaching Agreement," *Natural Resources Journal* 28(1):21–35 (1988).

GLOSSARY OF ACRONYMS

ADR   alternative dispute resolution

ARARs   applicable or relevant and appropriate requirements

BACT   best available demonstrated control technology

BAT   best available technology economically achievable

BCT   best conventional pollutant control technology

BPJ   best professional judgment

BPT   best practicable technology currently available

CERCLA   Comprehensive Environmental Response, Compensation, and Liability Act ("Superfund")

CWA   Clean Water Act

EDR   environmental dispute resolution

ESA   Endangered Species Act

FDF   fundamentally different factors

GAC   granular activated carbon

LDC   London Dumping Convention

MCL   maximum contaminant level

MCLG   maximum contaminant level goal

MPRSA   Marine Protection, Research, and Sanctuaries Act

NCP   National Contingency Plan

NEPA   National Environmental Policy Act

ONRWs   Outstanding National Resource Waters

POTW   publicly-owned treat-
       ment works

RCRA   Resource Conservation
       and Recovery Act

SDWA   Safe Drinking Water Act

SMCRA Surface Mining Control
       and Reclamation Act

SRFs   state revolving funds

TMDLs total maximum daily
       loads

UIC   underground injection
       control

USDW   underground source of
       drinking water

WSRA   Wild and Scenic Rivers
       Act

INDEX

Abandonment of water rights, 27, 34, 132, 157
Absolute ownership, 43–48, 63
Access rights, 22, 75, 118–121, 122–126, 131
Acid precipitation, 153–155, 254
Actions (legal), 1
Acts, 2
Actual diversion requirement, 35, 38
Adjudications, 3, 38–39, 157
Administrative fine, 3, 215, 217, 218, 245, 252, 261
Administrative regulations, 2, 3, 263
After-the-fact permit, 96
Amicus curiae, 265
Ancillary pollution, 210, 244
Anti-backsliding, 211
Antidegradation, 174–175, 240, 248, 258
Antiexportation statutes, 58–60
Appellate jurisdiction, 52
Appropriation. *See* Prior appropriation
Arbitrary and capricious, 36, 94
Area of origin statutes, 57–58
Artificial persons (corporations), 1

Artificial recharge, 19
Artificial uses, 23
Artificial water, 19–20
Artificial watercourses, 18, 21, 95, 123–126
Assimilative capacity, 170, 171, 186
Attainability, 175–176, 185

Basin management. *See* Watershed management and planning
Beach access, 115, 121
Beds and banks, 22, 118–121, 122–123
Beneficial use, 33–41, 157
Benefit-cost analysis, 81–82, 86, 98–99, 102, 178–179, 187
Best management practices. *See* Nonpoint source control
Better than best, 185, 187
Bioassay, 168, 170, 187, 209–210, 234, 236
Bonneville Power Administration, 88
Bubble concept, 188
Bureau of Land Management, 136–137

Bureau of Reclamation, 34,
    76–79, 82, 83, 87, 88, 106, 136,
    157, 159, 275
Bypass, 210

California Doctrine, 21, 33, 37, 45
Canada, 10, 55, 153–155, 220
Cases, 2, 4
Causation. See Proximate cause
Cause of action, 150
Certification of water quality,
    108, 131, 146, 212
Channelization, 87, 129
Citizen suits, 96, 146, 216–217,
    238, 261, 265
Civil penalty, 1, 215, 245, 252,
    261
Civil remedies, 1–2
Classes of water, 17–20, 39–40,
    41
Clean Lakes Program, 227
Coast Guard, 94, 238, 252, 254
Coastal zone management, 107,
    108, 143
Colorado doctrine, 33
Combined sewer overflows, 192,
    193, 194, 201, 206
Commerce Clause, 58–59, 73–74,
    89, 118, 144
Common enemy rule, 4, 61–63
Common law, 2, 4, 5, 22, 30, 64,
    65, 114, 138, 263–266
Compensatory damages, 1, 263
Compliance schedule, 209–210
Condemnation, 11, 24, 37, 38,
    63, 79, 90, 92, 93, 120,
    128–129, 157
Conditional water rights, 38
Congressional allocation, 52,
    54–55, 57
Conjunctive management, 17, 31,
    42, 44, 45–48, 59, 69, 225
Consent decree, 180–181, 204

Conservation of water, 6–7, 20,
    35–36, 41, 48, 53, 96, 99, 101,
    156–160, 200
Constitutions, 2, 3, 10, 11–12, 27,
    28, 30–31, 32, 65, 121
Construction grants program,
    193–201, 207, 220
Conventional pollutants, 179–180,
    183
Cooperative federalism, 212
Corps of Engineers, 51, 65, 79–85,
    86, 87, 88, 92, 93–96, 103–104,
    108, 124–126, 136, 143, 144–148,
    200, 235, 237–238, 275
Correlative rights (groundwater),
    43–48, 63
Corridor protection, 127–130,
    226, 227
Cost-sharing, 79–85, 100, 102
Courts (functions), 3
Criminal remedies, 1–2, 216, 238,
    245, 261
Customary law, 32, 33, 55, 121

Dam permits, 65, 94
Dam safety, 64–65, 280
Dams as pollutors, 177–178
Defendant, 1
Desalination, 101
Developed water, 20, 63, 152,
    159
Development of water resources,
    7–8, 9–40
Dewatering, 63–64
Diffused surface water, 17–18,
    21, 39, 61–63
Dilution of pollution, 108–109,
    179, 205
Direct discharge, 178
Discharge of a pollutant, 177
Discharge permits. See NPDES
    permit program
Distribution organizations, 66–69

Domestic sewage exclusion, 204
Domestic uses. *See* Natural uses
Drainage law, 4, 18, 61–65
Dredge and fill permits, 51, 95–96,
    124–126, 131, 143, 144–148
Dredged spoils, 233, 235,
    237–238, 280
Drinking water protection, 108,
    239–250, 258–262, 280
Duty of water, 36

Eastern diversion permits, 26–31,
    42, 43, 131
Effluent charges, 167
Effluent limitations, 168, 177–191,
    188–189, 209–210
Effluent limited stretches, 185
Eminent domain. *See* Condem-
    nation
Enabling act, 3, 80, 94
Endangered species, 88, 104–106,
    130, 131, 143, 146
Enforcement and enforceability,
    202, 207, 208, 210, 214–218, 220,
    224, 226, 230, 238, 245, 261
Environmental impact state-
    ments, 20, 81–82, 102–104,
    107, 132, 137, 142, 143, 145,
    147, 198–199, 209, 253
Environmental movement, 102
EPA. *See* United States Environ-
    mental Protection Agency
Equitable Apportionment, 52–55,
    230
Equity balancing, 150, 263
Erosion control, 85–87
Estuary protection, 191, 201, 227
Executive orders, 3–4, 99, 107–108

Facilities planning, 198–199
Fast lands, 75, 95
Federal common law of nuisance,
    221

Federal Emergency Management
    Agency (FEMA), 65, 139–140
Federal Energy Regulatory Com-
    mission (FERC), 90–93, 131, 275
Federal facilities, 212
Federal floor, 188, 212, 245
Federal outdoor recreation pro-
    grams, 135–137
Federal Power Commission. *See*
    Federal Energy Regulatory
    Commission
Federal regulatory water rights,
    51–52, 75, 131–133
Federal reserved rights, 49–51,
    130, 132
Federal-state conflicts, 15, 49–55, 92
Fee simple interest, 128
Fish and wildlife, 6, 50, 84–85,
    91, 96, 107, 130, 131, 133, 137,
    143–144, 172, 182, 233
Fish and Wildlife Service, 91,
    104, 105, 135, 136, 147
Fishable/swimmable waters, 173,
    174, 175, 178, 181, 184, 191, 258
Flood control, 77, 84, 145
Floodplains, 65, 107, 130,
    138–141, 142
Floodwaters, 18, 61, 64–65
Flowage easement, 65
Foreign water, 20, 40, 159
Foreshore. *See* Tidelands
Forest Service, 136–137
Forfeiture, 11, 29, 34–35, 36, 132,
    157
Forum shopping, 169, 171–172,
    185
Fundamental law (constitutions), 2
Fundamentally different factors,
    179

General permits, 96, 146–147, 192
Goals and policies of Clean
    Water Act, 171–173

Grandfather rights, 27–28, 154
Groundwater pollution, 177, 225,
    226, 230, 239–250, 252, 253–257,
    265–266
Groundwater rights, 4, 21, 42–48,
    54–55, 63–64, 115
Groundwater (state management
    systems), 45–48, 68–69, 115,
    157–158
Groundwater-surface water dis-
    tinction, 5–6, 17–20, 25, 42–43,
    61, 177
Guidelines, 3

Harbor construction and main-
    tenance, 83
Hazardous substances, 5, 130,
    159, 170, 204, 226, 241–247,
    249, 251–257, 265–266
Heat, 173, 181–182, 183, 280
Highway construction, 109
Historic preservation, 106–107
Hydroelectric power, 6, 7, 20, 35,
    77, 84, 88, 90–93, 109, 129,
    131, 280
Hydrology, 5, 17, 42, 43, 52, 81

Implied dedication, 121, 124
Incineration-at-sea, 236
Indian water rights, 49–51, 52,
    280
Indirect discharges. See Pretreat-
    ment
Industrial cost recovery, 199–200
Industrial dischargers, 170, 171,
    175, 177–189, 196, 223, 231,
    233, 244
Injunction, 1, 22, 215, 216, 245,
    263
Inland waterways, 83–84
Innovative and alternative waste
    treatment, 195–196, 198, 201,
    212

Instream flow protection. See
    Minimum flow protection
Integrity of a waterbody, 172,
    173
Interbasin transfers, 56–60
International waterways, 52
Interstate compacts, 52, 53–54,
    57, 157, 230, 270
Interstate water pollution,
    219–222, 229–230
Interstate waterways, 52–55
Iron triangle, 102
Irrigation, 6, 20, 23, 27, 35–37,
    66–69, 76–79, 84, 132, 145,
    156–158, 247, 249
Irrigation districts and organiza-
    tions, 34, 62, 66–69, 76–79, 86,
    90
Irrigation return flow, 19, 20,
    34–35, 36, 40, 156–157, 177,
    196
Issues, 6

Joint and several liability, 254
Junior appropriator, 33–35
Jurisdiction, 24, 52

Lakes, 85, 95, 118–121, 122–126,
    227, 264–265
Land and Water Conservation
    Fund, 127, 135–136, 143
Land use, 8, 165, 267–271
Law of capture, 18
Lawsuit, 1

Management of water resources,
    8–9, 10
Marine sanctuaries, 235
Marshes and swamps, 18, 61, 95
Mass allocation, 51
McCarran Amendment, 51
Mediation, 275–282
Mexico, 9, 10, 55

Minimum flow protection, 27, 31,
    35, 37, 41, 44, 50, 51, 75, 79,
    92, 115–117, 130, 131–133
Mitigation, 84–85, 147
Mixing zones, 169, 249
Monitoring, 210, 214, 216, 217,
    220, 226, 227, 234, 246
Municipal water supply, 20, 35,
    66–69, 77, 84, 132, 158–160

National Park Service, 121, 128,
    135–136
National Water Commission, 7
Natural flow doctrine, 22–23,
    61–63
Natural uses, 22, 23, 27, 35, 37
Navigability, 74, 94–96, 116,
    118–121, 122, 125, 177
Navigation, 6, 77, 92, 115, 118,
    122, 235
Navigation power, 73–75, 94–96
Navigation servitude, 65, 74–75,
    94, 125–127, 133
Negligence, 4, 5, 20, 63, 64, 65,
    90, 151, 253, 264
New sources, 178, 183, 187, 209,
    213
New water. See Developed water
No-injury rule, 34–35, 36, 40
Nonconventional/nontoxic pollu-
    tants, 173, 181, 183
Nonpoint source control, 143,
    145, 146, 169, 170, 173, 175,
    177, 186, 188, 201, 210, 211,
    220, 223–230, 233, 258
Nontransformational uses, 7–8,
    113–163
Notice of intention to divert,
    37–39
NPDES program, 170, 208–213,
    220, 233–234, 244, 252, 253
Nuclear Regulatory Commission
    (NRC), 89, 182

Nuisance, 4, 5, 90, 150, 264

Ocean disposal, 191, 233–238
Oil pollution, 236, 251–253
Ordinances, 2
Organic Act, 3
Original jurisdiction, 52
Outstanding National Resource
    Waters, 174–175

Parameters, 167, 179
Parks management, 130, 174–175
Percolating groundwater, 18–19,
    39
Performance standards, 249
Permits and common law, 65,
    264–266
Place-of-use restrictions, 21–25,
    43–48, 57
Plaintiff, 1
Planning (water resources),
    97–100, 115, 224–230, 243, 248
Point sources, 175, 177–178
Pollution of water, 4, 7, 171–173
Precedent, 4
Preferred uses, 23, 28, 29, 36–37,
    41
Prescriptive rights, 24, 35, 120,
    121, 124
Pretreatment, 159, 178, 187, 190,
    191, 199–200, 202, 203–207,
    212, 218, 237, 264
Principles and standards, 98–100
Prior appropriation, 3, 21, 22, 28,
    29, 32–41, 43–48, 51, 52–53, 63,
    116–117, 132–133, 150, 152,
    157, 158–159, 186, 196–197
Priority pollutants, 181, 204
Private property, 10–12
Privatization, 201
Prohibited pollutants, 182, 193,
    235
Promulgation of regulations, 3

Protest proceedings, 37
Proximate cause, 4, 138, 151–152, 263–264
Public access. *See* Access
Public interest review, 96, 146–148
Public policy, 8–9
Public trust doctrine, 11, 90, 114–117, 118, 119, 121, 133, 143
Public use of waterbodies, 118, 121
Publicly-owned treatment works (POTWs), 159, 170, 175, 180, 186, 190–192, 193–202, 218, 223, 227, 232, 233, 237, 261, 265, 280
Pueblo rights, 38–39, 159

Reasonable beneficial use, 36, 37, 157
Reasonable use rule (drainage), 4, 61–63
Reasonable use rule (groundwater), 43–48, 63
Reasonable use rule (surface water), 21–25, 43–44
Reclamation law, 76–79
Recreation (water based), 6, 8, 9, 20, 22, 35, 37, 50, 75, 77, 84, 91, 92, 115, 118–121, 122–123, 127–134, 135–137, 158, 172, 174, 201–202, 233, 280
Regional water management, 26, 142, 219–222, 224–230, 267–271
Registration of diversions, 11
Regulatory strategy, 167
Relation back, 38
Remedies, 1
Removal allowances, 204–207
Research, 100–101, 186, 220, 240
Restatement of Torts rule (groundwater), 44
Restoration order, 96
Rights, 1, 10–12

Right to recapture, 48, 57
Riparian, 21
Riparian rights, 4, 21–25, 37, 39, 53, 65, 75, 123, 131, 150, 152, 264
Riparian system (riparianism), 21–25, 32–35
River basin commissions, 98–100
Rotation plans, 39
Rule-making. *See* Administrative regulations

Saline intrusion, 9, 30, 248
Saline water conversion, 262
Salvage water, 20, 35–36, 40, 157, 158
Savings clause, 264
Sawlog test, 74, 94, 119
Scenic/aesthetic uses, 6, 7, 35, 50, 115–117, 133, 137
Scenic easements, 128, 130
Science and law, 4–6, 7, 43, 151–152, 234, 264
Seepage, 19–20
Senior appropriator, 33–35
Septic systems, 194, 226, 232, 241, 247, 249
Sewage sludge, 190, 191, 203–207, 234–237, 241, 264, 280
Sewer bans, 216
Slippage, 140
Small watersheds, 85–87
Soil Conservation Service (SCS), 85–87, 136
Source of title test, 22
Sovereign immunity, 63, 65, 138, 152, 265
Spill control, 226, 233, 251–257
Spite pumping, 43
Springs, 18, 39, 61
Standing to sue, 10, 57, 216, 256, 263
State certification. *See* Certification of water quality

State constitutions, 2, 3, 11,
    15–16, 39
State engineer, 3, 37–39, 62
State powers (water diversion),
    15–16
State revolving funds (SRFs),
    193, 201–202, 227
State supreme courts, 2, 4, 5
Statutes, 2–3, 4, 32
Stop work order, 96
Storage of water, 20, 35, 40, 108,
    132
Stormwater management, 61–63,
    66, 178, 190–192, 223–230, 249
Strict liability, 64, 151, 253, 254,
    257
Subflows (underflows), 18–19
Subsidence, 63–64, 265
Support, duty of, 63–64

Tailwater. *See also* Irrigation
    return flow, 20
Takings, 2, 11–12, 15, 28, 30–31,
    65, 90, 117, 125, 140–141,
    142–143, 264
Tax credits, 130, 144, 201
Technical standards, 249
Technology-based approach, 167,
    170, 175, 178–183, 184, 185,
    186, 189, 190, 204–205, 223,
    234
Technology-based effluent limita-
    tions. *See* Technology-based
    approach
Tennessee Valley Authority
    (TVA), 87–88, 271
Third-party recoveries, 253, 255,
    256–257, 264
Tidelands, 95, 114, 115, 118, 121
Torts, 63
Total maximum daily loads
    (TMDLs). *See* Wasteload allo-
    cations

Toxic pollutants, 146, 169,
    180–181, 183, 187–188, 190,
    199, 204, 210, 217, 223, 232,
    236, 253, 258–262, 265
Trading effluent limitations, 188
Training grants, 200
Transfer of water diversion
    rights, 11, 24, 34–35, 40–41,
    68, 157, 158
Transformational uses. *See also*
    Nontransformational uses,
    7–8, 29
Trespass, 150–151, 264
Tributary groundwater, 19,
    46–47, 63

Underground injection, 241–242,
    258
Underground streams, 18–19
United States Constitution, 2–3,
    10, 11, 15, 53–54, 58–59, 65,
    73, 95
United States Geological Survey
    (USGS), 100–101
United States Environmental Pro-
    tection Agency (EPA), 3, 90,
    104, 109, 144–148, 154, 167–261,
    275
United States Supreme Court, 2,
    49–50, 52–55, 57, 58–59, 73, 74,
    79, 88, 105, 114–115, 124–126,
    133, 160, 179, 215, 221–222,
    230
Unity of title test, 22
Upsets, 210
Use attainability analysis,
    175–176, 184, 189
User charges, 199, 200, 202
Uses of water, 6–8
Usufructuary right, 11

Value disagreements, 277, 280
Veto, 2–3, 100

Waste disposal, 6, 7–8
Wasteload allocations, 168, 171,
    185, 186, 211, 220, 231, 234
Waste of water, 35–36, 37, 41,
    43–48, 53, 156
Wastewater, 19–20, 40, 158, 159
Wastewater treatment, 66
Water banking, 157, 158
Water codes, 37–39
Watercourse, 18
Water courts, 38
Watermasters, 39
Water quality-based approach,
    167–170, 173, 178, 180, 182,
    183–189, 191, 211, 234
Water quality criteria. See Water
    quality standards
Water quality-based effluent
    limitations. See Water quality
    standards
Water quality law, 9, 130, 142,
    157, 165–271, 280
Water quality-limited stretches,
    185, 186, 189, 190, 194, 231
Water quality management plan-
    ning, 198–199, 224–230,
    231–232

Water quality standards, 167–170,
    171, 178, 181, 183–187, 220
Water quantity law, 9, 11, 280
Water Resources Council, 98–100
Water rights, 10–12
Watershed management and
    planning, 85–88, 91, 97–100,
    128, 130, 137, 226, 227, 229,
    260, 267–271, 280
Waters of the United States,
    95–96, 177
Weather modification, 149–152,
    159
Well-spacing statutes, 8
Wetlands, 107–108, 115, 130,
    142–148, 177, 249
Wharfing out, 22, 75
Wild and scenic rivers, 93, 106,
    27–130, 131, 142, 143, 146, 268
Wilderness areas, 108
Win-Win result, 278
Winters Doctrine. See Federal
    reserved rights

Zero-discharge, 170, 173, 178, 186
Zoning, 10, 11, 128, 129, 130,
    169, 249, 268

DATE DUE

OCT 2 5 1989			
DEC 1 6 1995			
GAYLORD			PRINTED IN U.S.A